International Vocational Education in Key Technologies for Guangzhou Metro Construction, Operation, and Maintenance

Table of Contents

Part I: Guangzhou Metro and the Internationalization of China's Rail Transit

As the opening chapter of this book, this section aims to systematically and thoroughly examine the epic development of China's urban rail transit, from its arduous beginnings to its current position of global leadership. By delineating its developmental phases, quantifying its current scale, deconstructing its technological systems, analyzing its policies and standards, and forecasting future trends, it constructs a comprehensive and profound macro-level understanding of China's urban rail industry for the reader. This chapter focuses on how China's urban rail sector, through large-scale construction practice, has forged a unique system of key technologies in planning, design, construction, operation, and maintenance, ultimately transforming its role from a "technology follower" to a "market leader" and even a "standard setter." All this lays a solid historical and industrial foundation for subsequent chapters focusing on Guangzhou Metro's specific technical practices and its international vocational education exports.

Chapter 1： Overview of Urban Rail Transit Development in China

1. Introduction: Urban Arteries and the Pulse of the Era

Urban rail transit, often metaphorically referred to as the "arteries" and "skeletal system" of modern cities, is far more than just a high-capacity, efficient, low-pollution, and safe public transportation mode.

It stands as a core indicator of a city's modernization level, economic vitality, governance capability, and overall competitiveness.It profoundly reshapes urban spatial patterns, guides the optimal allocation of land resources, and transforms citizens' lifestyles, serving as a crucial engine for sustainable urban development.

For China, undergoing the world's largest and fastest urbanization process, developing URT is imperative to alleviate "urban diseases" such as traffic congestion and air pollution in large cities.

It is also a natural choice to optimize urban spatial structures, promote regional coordinated development, and implement national strategies for green and low-carbon development. China's UR development history is a condensed reflection of its reform and opening-up, technological innovation, and industrial upgrading. From initial technological admiration and comprehensive importation to digestion, absorption, integrated innovation, and now full-scale independent innovation and global leadership, China's URT industry has traversed a path in decades that took developed countries over a century, establishing the world's most extensive, complex, and modern URT network. Understanding this magnificent journey is essential for comprehending Guangzhou Metro's technological achievements, industry status, and its international vocational education mission.

2 Development History: From Humble Beginnings to Global Leadership

China's URT development has been non-linear and eventful, clearly divided into several strategic phases, each marked by distinct era-specific characteristics and development themes.

2.1 Infancy and Exploration Phase (1960s-1990s)

Context and Motivation: URT construction during this period had strong political and military overtones, with "wartime preparedness as the primary goal, supplemented by civilian use" as the core guiding principle.

China's economic foundation was weak, and technological reserves were insufficient.

Milestone - Beijing Metro Line 1 (1965-1969): As China's first metro line, its construction was carried out under extremely challenging conditions.

Utilizing open-cut excavation (shallow burial method), the 23.6-kilometer line connected Beijing Station and Pingguoyuan in the western suburbs. Its vehicles and equipment were mostly domestically produced imitations, with a significant technological gap compared to international advanced levels at the time. Nevertheless, it marked China's breakthrough in the URT field, nurturing the country's first generation of URT design and construction professionals and accumulating invaluable early experience.

Figure 1-1 Beijing Subway Line 1 Route Map

Slow Development: Subsequently, until the early stages of reform and opening-up, only Beijing and Tianjin had metro systems in the 1980s, with short line lengths, outdated technology, and weak passenger transport functions.

URT was far from becoming the mainstream of urban public transportation.

2.2 Take-off and Popularization Phase (1990s-Early 21st Century)

Context and Motivation: With the deepening of reform and opening-up, exemplified by the development of Shanghai Pudong, China's economy entered a fast lane.

Speculative cities experienced rapid population growth and economic activity, leading to soaring transportation demands and increasingly severe ground traffic congestion. The economic and social need for modern URT became urgent.

Milestone - Shanghai Metro Line 1 (1993): This marked the true starting point of China's modern URT.

Funded by a German government loan totaling approximately 6 billion yuan, it comprehensively introduced advanced German technologies, including vehicles, signaling, power supply, and shield tunneling equipment, along with management expertise. For the first time, shield tunneling technology was employed to traverse Shanghai's soft soil layers, and a modern operational management system was established. The construction of Shanghai Metro Line 1 served as a "Yellow Military Academy," systematically cultivating China's first generation of URT planning, construction, and management professionals proficient in international standards and modern technologies.

Policy Breakthrough and Regulation: In response to the emerging "metro fever" in some cities, the State Council issued the "Notice on Suspending the Approval of Urban Underground Rapid Transit Projects" in 1995, temporarily halting approvals but also signaling the state's macro-level planning and standardized management of this strategic infrastructure.

Figure 1-2 Shanghai Metro Line 1 Route Map

Multi-City Initiation: With increasing economic strength, Guangzhou, Shenzhen, and other southern cities at the forefront of reform and opening-up joined the construction ranks.

Guangzhou Metro Line 1 (1997) similarly adopted a significant amount of imported equipment and financing models, incorporating more urban planning concepts into its construction. This period was characterized by "exchanging market access for technology," rapidly narrowing the gap with international advanced levels through international cooperation and high-standard imports while nurturing the domestic industrial chain's infancy.

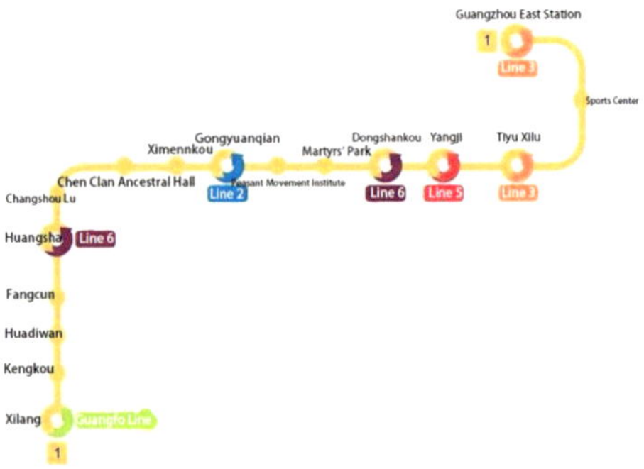

Figure 1-3 Guangzhou Metro Line 1 Route Map

2.3 Period of Rapid Development and Network Expansion (Early 21st Century – Mid-2010s)

(1) Background and Driving Factors: China's accession to the WTO unleashed the benefits of economic globalization, accelerated urbanization, and strengthened national financial capacity. In response to the 2008 global financial crisis, the state introduced a series of economic stimulus policies, making infrastructure investment a key tool for stabilizing economic growth.

(2) Policy Support and Investment-Driven Growth: The state relaxed approval conditions for urban rail transit projects, leading to the approval of metro system constructions in a large number of provincial capitals and cities with independent planning status, such as Shenyang, Chengdu, Wuhan, Nanjing, and Xi'an. China's urban rail transit entered a phase of "vigorous growth" with explosive expansion. Cities invested unprecedentedly, shifting their focus from single-line construction to network planning and layout. The complex design and construction of large transfer hubs became a central technical challenge.

(3) Localization Strategy and Technological Innovation: The state's "70% localization rate" policy significantly stimulated and promoted the rise of domestic equipment manufacturing. CRRC's manufacturing plants achieved breakthroughs in vehicle production through technical cooperation and independent research, while companies like China Railway Signal & Communication Corp (CRSC) made critical strides in the localization of signaling systems (CBTC). Domestic enterprises moved beyond imitation, entering a stage of integrated innovation and partial original innovation, with cost advantages gradually becoming apparent.

2.4 New Phase of High-Quality Development and Intelligent Transformation (Mid-2010s – Present)

(1) World's Largest Scale: China's total urban rail operating mileage has experienced exponential growth, firmly securing its position as the world's largest network, with the leading advantage continuously expanding. Network coverage has extended from central urban areas to surrounding new towns, suburban airports, and high-speed rail stations. Multiple systems, such as regional (suburban) railways,

tramways, and medium-to-low speed maglev, have achieved coordinated development, manifesting a trend toward "integrated multi-network systems."

(2) Connotative Growth: National authorities have strengthened macro-control, placing greater emphasis on the socio-economic benefits and fiscal sustainability of projects. The development focus has shifted from "competing on investment, mileage, and speed" to "competing on quality, efficiency, and service." The industry is increasingly prioritizing operational safety, passenger experience, Transit-Oriented Development (TOD) benefits, and green, low-carbon development.

(3) Intelligent Leadership: New-generation information technologies such as cloud computing, big data, the Internet of Things, artificial intelligence, 5G, and BIM are deeply integrated with urban rail transit. GoA4-level Fully Automatic Operation (FAO) systems have been deployed on multiple new lines; intelligent maintenance platforms are widely adopted; smart stations, intelligent security checks, and contactless payments enhance service quality. China's urban rail transit has begun transitioning from "running alongside" to "leading" in applied technological innovation, contributing "Chinese wisdom" to global urban rail development.

3. Status and Scale: A Grand Landscape of Remarkable Achievements

As of the end of 2025, the achievements of China's urban rail transit system can be outlined by a series of detailed data, revealing its grand scale:

(1) Number of Operating Cities: Urban rail transit systems are now in operation in more than 55 cities across China. From megacities to medium-sized cities, urban rail has become a "standard feature" of Chinese urban modernization.

(2) Operating Mileage and Networks: The total operating mileage has surpassed the 10,000-kilometer mark, firmly ranking first in the world. Specifically, the operating mileage in both Shanghai and Beijing exceeds 800 kilometers, while cities like Guangzhou, Chengdu, Shenzhen, and Wuhan have each surpassed 500 kilometers, forming the world's largest cluster of metro networks.

Top 10 Cities by Subway Operating Mileage

Rank	City	Operating Lines	Operating Mileage
1	Beijing	29	879 km
2	Shanghai	22	872 km
3	Guangzhou	19	694 km
4	Chengdu	15	634 km
5	Shenzhen	18	595 km
6	Wuhan	15	561 km
7	Hangzhou	12	516 km
8	Chongqing	11	495 km
9	Nanjing	15	483 km
10	Xi'an	11	385 km

Figure 1-4 Top 10 Cities in China by Urban Rail Operating Mileage

(3) Passenger Volume and Service: The annual passenger volume exceeds 29 billion. The average daily passenger volume in super-sized cities like Beijing, Shanghai, Guangzhou, and Shenzhen surpasses 10 million, with peak daily figures repeatedly hitting new records. On average, urban rail carries over 50% of all public transport trips in cities, exceeding 70% in core urban areas, solidifying its role as the undeniable backbone of urban passenger transport.

(4) Diversity and Systematization of Technology Models: The system boasts a comprehensive range of technology models, including metros, light rail, monorail, regional express rail, modern trams, maglev transit, and APM (Automated People Mover). This has formed a diversified, multi-level, and complementary rail transit system adaptable to different city sizes, passenger flow demands, and fiscal capacities.

(5) Industrial Chain Completeness and Competitiveness: China has established the world's most complete, large-scale, and competitive industrial chain for urban rail.

This chain encompasses macro policy research, planning and design consultancy, engineering procurement and construction (e.g., "China Railway Construction Corporation", "China Railway Group Limited"), equipment manufacturing (e.g., "CRRC", "China Railway Signal & Communication Corporation"), operations management, and value-added services (e.g., TOD development, commercial advertising). Chinese urban rail enterprises not only meet domestic demand but are also aggressively expanding into the international market, leveraging their advantages in technology, cost, and experience.

4. Evolution, Breakthroughs, and Innovation in the Key Technology System

The fundamental driver behind the leapfrog development of China's urban rail transit has been the continuous innovation and systemic breakthroughs in key technologies. The vast domestic market has provided a uniquely broad stage for technological practice and iterative innovation.

4.1 Planning and Design Technology: From Extensive to Intensive, from Isolated to Integrated

(1) Network-wide Collaborative Planning: The planning philosophy has evolved from isolated single-line planning to integrated, multi-level, multi-modal "One-Network" collaborative planning. Big data and simulation technologies are used for passenger flow forecasting and network assessment, optimizing route alignment, station layout, and transfer schemes. Emphasis is placed on seamless integration with national rail trunk lines, intercity railways, airports, and highway passenger hubs, building a comprehensive "Four-Network Integration" transport system.

(2) Deepening and Innovation of TOD: Tightly integrating rail transit construction with comprehensive land development along the lines, urban function enhancement, and industrial introduction, a mature "Rail + Property" development model with Chinese characteristics has been formed. Through over-track development and underground space interconnection, land value capture is used to subsidize rail transit construction and operation, solving the funding challenge for sustainable development. Practices by Shenzhen Metro and MTR (Shenzhen) have become global

exemplars.

(3) Green and Resilient Design: Concepts such as green, low-carbon, sponge city, and disaster prevention/mitigation are fully integrated into the planning and design stages. Application of technologies like "energy-saving gradient" line design, photovoltaic power generation, heat pump systems, energy feedback devices, and green building materials is promoted, building a green urban rail system throughout its lifecycle. Simultaneously, disaster resilience capabilities such as flood control and seismic resistance for stations and lines are enhanced.

4.2 Construction Technology: Challenging Engineering Limits, Leading Method Revolution

(1) Leap in Tunnel Construction Technology: The Tunnel Boring Machine (TBM) method has become the absolute mainstream for tunnel construction in complex urban environments. China has not only completely overcome worldwide challenges like soft soil, sandy cobble strata, composite strata, and high water pressure but has also achieved "Chinese Creation" in TBMs – domestically produced TBMs (e.g., from China Railway Engineering Equipment Group, China Railway Construction Heavy Industry) now hold over 90% of the domestic market share and are exported to high-end markets like Paris and Singapore. High-end equipment and technologies like large-diameter TBMs and dual-mode TBMs have reached world-leading levels.

(2) Deep Excavation and Underground Mining Technology: In city centers dense with buildings and utilities, deep excavation engineering techniques have become increasingly sophisticated. Support technologies like diaphragm walls, TRD (Trench Cutting Re-mixing Deep Wall) method, and ultra-deep triple-axis mixing piles, along with precise dewatering and settlement control technologies, are widely used. Methods like the shallow buried tunneling method and the Pile-Beam-Arch (PBA) method effectively replace cut-and-cover in suitable ground conditions, minimizing disruption to the urban surface.

(3) Popularization of Prefabricated Construction (PC Method): Factory prefabrication and on-site assembly are widely promoted for station structures,

segment rings, track beds, MEP (Mechanical, Electrical, and Plumbing) pipelines, and even station entrances/exits. This not only significantly improves project quality, construction efficiency, and safety but also markedly reduces on-site noise and dust pollution, making it a model for green construction.

4.3 Equipment Manufacturing Technology: From Localization to Self-Sufficiency, from Follower to Leader

(1) Comprehensive Breakthroughs in Rolling Stock Technology: CRRC can independently research, develop, and manufacture a series of product platforms covering different car types (A/L/B/As, etc.) and speeds ranging from 80-160 km/h (and even higher-speed regional trains and high-speed maglev). It has achieved international leading levels in permanent magnet synchronous traction systems, lightweight technologies like carbon fiber composites, noise reduction technology, and smart passenger information systems. Chinese metro trains are sold globally due to their high reliability, cost-effectiveness, and good adaptability.

(2) Self-Sufficiency and Control in Signaling Systems: Successfully breaking the long-term monopoly of foreign giants (Siemens, Alstom, Thales), CBTC systems with completely independent intellectual property rights, developed by companies like China Railway Signal & Communication Corp (CRSC) and Traffic Control Technology Co (TCT), have become mainstream in the market. Further evolution towards more efficient and flexible next-generation signaling systems like VBTC (Vehicle-to-Vehicle Based BTC) and virtual coupling is underway. GoA4-grade Fully Automatic Operation (FAO) system technology is mature and has been deployed in Shanghai, Beijing, Guangzhou, and elsewhere.

(3) Intelligent Upgrading of the Entire Industrial Chain: Smart AFC systems support various payment methods like QR codes and biometrics; Intelligent Integrated Supervisory Control Systems (ISCS) achieve deep integration and linkage of subsystems; smart platform screen doors, intelligent lighting, and intelligent environmental control systems are widely used, comprehensively enhancing system intelligence.

4.4 Operation and Maintenance Technology: From Experience-Driven to

Data-Intelligence-Driven

(1) Networked Operation Dispatching: Faced with ultra-large-scale networks spanning thousands of kilometers, powerful Network Operation Control Center (NOCC) systems have been developed, enabling real-time monitoring, intelligent analysis, and coordinated dispatching of the entire network's operational status and passenger flow data. China's capability in managing ultra-high passenger volumes is world-class.

(2) Building an Intelligent Maintenance System: This is a current hotspot for technological innovation. Utilizing IoT sensors, train-to-ground wireless communication (5G), big data analytics, and AI algorithms, systems for equipment condition monitoring, data fusion, and Prognostics and Health Management (PHM) are being built. This shifts maintenance from traditional "scheduled maintenance" and "breakdown maintenance" to condition-based "predictive maintenance," greatly enhancing equipment reliability, reducing maintenance costs and failure rates, and ensuring operational safety.

(3) Comprehensive Safety Assurance System: A comprehensive, multi-dimensional safety assurance system and emergency response plans have been established, covering construction safety, operational safety, equipment safety, fire safety, and counter-terrorism/security. Technologies like video intelligent analytics and intelligent security screening are used to improve the precision and efficiency of safety prevention and control.

5. Drive and Standard System Construction: The Strategic Power of Top-Level Design

5.1 Strong Guidance from National Strategy and Policy

The rapid development of China's urban rail transit is inseparable from strategic planning and precise policy regulation at the national level. From early-stage approval controls to mitigate risks, to later encouragement for development to stimulate the economy, and now to standardized management emphasizing high-quality development and risk prevention, a series of top-level design documents—such as the

Outline for Building a Leading Transportation Country, the Outline of the National Comprehensive Three-Dimensional Transportation Network Plan, and the Guiding Opinions on Further Improving Railway Planning and Construction Work and their guiding spirit for urban rail—have charted the development direction, set access thresholds, and standardized the development order for the industry. Financial inputs, local government bonds, and innovations in PPP models have also provided diversified funding channels for the massive construction investments.

5.2 The Formation and Internationalization of the Chinese Standard System

In the early stages of development, China widely adopted European (EN), International Electrotechnical Commission (IEC), and other standards. With the enhancement of technical capabilities and the accumulation of extensive practical experience, China began systematically building its own technical standard system. Led by the National Standardization Committee, the Ministry of Housing and Urban-Rural Development, the Ministry of Transport, and others, hundreds of Chinese National Standards (GB) and industry standards (such as CJJ/T, TB, etc.) covering design, construction, acceptance, operation, and equipment manufacturing have been formulated. This "Chinese Standard" system is rooted in China's complex geological conditions, huge passenger flow loads, and high-intensity operational demands, having been tested in the most rigorous practice; therefore, its technical indicators are often more advanced, economical, and applicable. Today, Chinese urban rail standards are being applied and promoted in projects across Asia, Africa, South America, and elsewhere, following the "Belt and Road" Initiative and the "Going Global" strategy of Chinese enterprises. This shift from being a "standard follower" in the past to a "standard setter" and "exporter" is a key indicator of China's transition from a large urban rail country to a strong one.

6 Challenges and Future Trends: Towards a Sustainable and Intelligent Future

While tremendous achievements have been made, China's urban rail also faces new challenges and development opportunities. Future development will show the following trends:

6.1 Current Core Challenges

- **Financial Sustainability Challenge: High** construction investment and operational costs place heavy debt pressure and operational subsidy burdens on local governments and enterprises. "How to make money" rather than "continuously burning money" is a common industry problem, making the exploration of diversified profit models and TOD feedback mechanisms crucial.

- **Unbalanced Development and Precise Supply: Networks** in core areas of first- and second tier cities are maturing, while suburban lines in some cities and systems in third- and fourth-tier cities face issues of insufficient passenger flow intensity and poor economic returns. The future requires more precise matching of passenger flow demand and the development of suitable medium- and low-capacity transit systems, avoiding the blind pursuit of metro systems.

- **Long-term and Complex Safety Risks:** As network scale expands, equipment ages, and extreme weather events increase, the pressure for risk control in safe operation continues to grow. New risks such as cybersecurity and data security are also becoming increasingly prominent.

- **Structural Shortage of Talent:** There is still a shortage of high-level, interdisciplinary technical and management talents, especially experts proficient in smart technologies and international business.

6.2 Future Development Trends

- **Deep Evolution of Smart Urban Rail**: Building the "Urban Rail Cloud" and big data platforms will form the digital foundation, promoting the deep application of AI across all scenarios including intelligent dispatch, intelligent service, intelligent maintenance, and intelligent safety, achieving global optimization and autonomous evolution of the urban rail system.

- **Comprehensive Transition to Green and Low-Carbon:** Fully implement the "Dual Carbon" strategy, create near-zero carbon stations, energy-saving train systems, large-scale application of renewable energy such as photovoltaic and hydrogen power, establish carbon footprint monitoring and management systems, and build a green urban rail system throughout its lifecycle.

- **Deepened Implementation of "Four-Network Integration":** Break down management system and technical barriers to truly achieve deep integration of urban rail transit, regional (suburban) railways, intercity railways, and national trunk railways in planning, construction, operation, management, and ticketing, providing passengers with a convenient experience of "one-stop ticketing, one-code passage, and seamless transfer".
- **Upgraded International Expansion:** Shift from past single-product (rolling stock) exports and EPC project contracting towards the export of integrated industrial chain solutions encompassing "Technology + Standards + Equipment + Management + Operation + Training", moving from "Going Out" to "Going In" (localized cultivation) and "Going Up" (participating in high-end market competition).
- **Resilience and Security as Standard:** Enhance the system's capability to predict, warn against, protect from, and quickly recover from sudden risks such as public health incidents, natural disasters, and cyber attacks, building a more resilient urban rail system.

Chapter 2: Technological Innovation and International Practices of Guangzhou Metro

This chapter aims to provide a comprehensive exposition of the remarkable technological innovation journey and highly effective international practices of Guangzhou Metro Group Co., Ltd. (hereinafter referred to as "Guangzhou Metro"), a leader in China's urban rail transit industry. The development history of Guangzhou Metro is an inspiring epic, evolving from "Introduction, Absorption, and Digestion" to "Integrated Innovation," then to "Independent Innovation," and ultimately achieving "Technological Leadership." This chapter will systematically analyze the core technological systems honed by Guangzhou Metro in addressing challenges such as complex geological conditions, ultra-high-intensity operations, and intelligent development. These encompass key areas including planning and construction in complex environments, advanced equipment manufacturing, and intelligent operations and maintenance. Furthermore, this chapter will delve into how Guangzhou Metro has translated its technological accumulation and management experience into a competitive advantage internationally. It explores how, through various models such as high-end consulting, EPC contracting, operation management, technical cooperation, and vocational education and training, Guangzhou Metro has gone global. In serving the "Belt and Road" Initiative, it provides "Guangzhou Solutions" and "Chinese Expertise" to the global urban rail industry, thereby laying a solid foundation for its mission in international vocational education.

2.1 Introduction: From Urban Commuter Backbone to Global Industry Benchmark

Born at the forefront of China's reform and opening-up, Guangzhou Metro has seen its development trajectory grow in lockstep with the rise of the Guangdong-Hong

Kong-Macao Greater Bay Area since the opening of its Line 1 in 1997. Evolving from an initial 18.5-kilometer line to a massive network now managing over 857 kilometers of operational mileage, with an average daily passenger flow exceeding ten million, Guangzhou Metro has transcended its role as the city's main transportation artery. It has firmly established itself as a world-class industry benchmark, consistently ranking first globally for operational safety performance and on-time rates for many years.

However, Guangzhou Metro's ambitions extend far beyond this. It serves as a continuously driven engine of technological innovation and a platform for international practice. Throughout its nearly three-decade development history, Guangzhou Metro has confronted world-class engineering challenges, including karst geology, fault zones, soft soil foundations in the Pearl River Delta, river crossings beneath the Pearl River, and construction in densely built urban areas. It has also managed extreme passenger surges during events like the Spring Festival travel period and the Canton Fair. In the process of solving these complex problems, Guangzhou Metro has transformed from an initial technology follower into a leading integrated innovator and a driver of original innovation. This chapter will provide an in-depth analysis of this transformation and demonstrate how it has translated its core competencies into global influence.

Table 2-1 Guangzhou Metro Mileage Changes Over the Years

No.	Year	Mileage Status
1	1997	The first line (Line 1, Xilang to Huangsha section) was opened, with an operational mileage of 5.4 kilometers.
2	2009	The network scale reached 150 kilometers, covering 88 stations.
3	2015	Operational mileage exceeded 300 kilometers.
4	2022	Operational mileage reached 607.6 kilometers.
5	2023	Operational mileage increased to 653.4 kilometers, ranking fourth globally.
6	June 2025	Operational mileage reached 751.1 kilometers, ranking third in China.

2.2 Development Journey: Three Strategic Leaps from 'Following' to 'Leading'

The technological innovation path of Guangzhou Metro can be distinctly divided into three strategic phases, achieving the leap from catching up to leading.

(1) Era 1.0: Introduction, Absorption, and Re-innovation (1992-1999)

•**Landmark: Line** 1 Construction.

•**Characteristics: Funded** partially by foreign loans, it comprehensively introduced German equipment, technology, and management standards, starting from a high baseline. The core task of this phase was 'learning' and 'digestion,' cultivating the first generation of modern urban rail talent and establishing a preliminary operation management system. Although the construction cost was high, it laid a solid foundation for subsequent development.

(2) Era 2.0: Localization and Re-innovation (2000-2010)

•**Landmark: Line** 2 Construction.

•**Characteristics: As** China's first national rail transit localization demonstration project, it responded to the national requirement that the localization rate of mechanical and electrical equipment must not be less than 70%. Guangzhou Metro pushed itself further, demanding that Line 2's overall standard not be lower than Line 1's.

Breakthrough Achievements:

• **Platform Screen Doors (PSD): Despite** pressure, Guangzhou Metro collaborated with domestic enterprises to successfully develop and apply localized PSD systems. This enhanced operational safety, energy efficiency, and environmental protection, later becoming a national standard.

• **Shield Tunneling Technology Localization:** Leveraging experience from Line 1, while core components for Line 2's shield tunneling were imported, the design and construction were completed independently. Guangzhou Metro established theories and methods for shield tunnel construction, laying the technical and talent foundation for the widespread adoption of shield technology in China.

• **Outcome: Line** 2 was ultimately completed in half the time, with the cost per kilometer reduced from 660 million yuan (Line 1) to 480 million yuan. Its technological integration and application won the Second Prize of the National

Science and Technology Progress Award in 2006.

(3) Era 3.0: Independent Innovation and Smart Leadership (2011-Present)

Landmarks: Large-scale network construction (e.g., Line 3, Line 14) and exploration of smart metro (e.g., Line 18, Line 22). Guangzhou Metro entered an explosive period of independent innovation.

Representative Innovations:

• **High-Speed Regional Express Lines: Line** 3 pioneered the use of 120 km/h express trains with limited stops in China. Lines 18 & 22 were the first to adopt 160 km/h Type D regional trains, achieving high-frequency, metro-like operation.

• **National Engineering Laboratory: In** 2016, Guangzhou Metro took the lead in establishing the "National Engineering Laboratory for System Safety and Operation Assurance of Urban Rail Transit", becoming a crucial platform for national-level scientific and technological innovation.

• **Smart Metro: Leading** national projects like the 863 Program and National Key R&D Programs (e.g., 5G projects), Guangzhou Metro launched its independently developed Suiteng OS intelligent operating system. It applies cutting-edge technologies such as Fully Automatic Operation (FAO), Digital Twin, and Intelligent Maintenance, redefining the future paradigm of urban rail development.

2.3 Core Technological Innovation System

Guangzhou Metro's technological innovation has evolved into a systematic, multi-dimensional core technology system covering the entire lifecycle.

2.3.1 Planning and Construction Technology in Complex Environments

• **Karst Treatment Technology:** Established a mature comprehensive technical system of "Detection-Analysis-Treatment-Evaluation," setting a benchmark for construction technology in karst regions domestically and internationally.

• **Shield Tunneling Technology in Composite Strata:** Innovated in shield machine selection, parameter control, and soil conditioning technology, overcoming the challenges of tunneling through strata with "upper-soft and lower-hard" compositions. Its experience has become an industry model.

• **Micro-Disturbance Control Technology in Densely Built Areas:** Developed high-precision monitoring and intelligent control technologies to minimize the impact of construction on the urban environment.

• **High-Efficiency and Energy-Saving Construction Methods:** Promoted the large-scale application of Prefabricated Construction (PC) technology, enhancing project quality, efficiency, and green construction standards.

2.3.2 Advanced Equipment and Intelligent Manufacturing Technology

• **High-End Train Platform:** Co-developed the 160 km/h Type D regional train, deployed on Lines 18 and 22, filling the capacity gap between conventional metro and intercity services.

• **Fully Automatic Operation (FAO) System:** Actively implemented and applied the GoA4-grade FAO system, achieving fully automated train operation.

• Tencent **OS Intelligent Operating System:** A pioneering system in China, it enables functions like contactless fare gating, intelligent passenger flow management, one-click station startup/shutdown, and intelligent equipment maintenance. Its application has reduced the ratio of maintenance personnel by 16% and cut manual patrol inspection time from 1 hour to 10 minutes.

• **5G Integration and Application:** Leading national 5G R&D projects, it developed the "5G Train-borne Edge Gateway Equipment (TAU)" and the "Metro Intelligent Decision-Making Platform," enhancing train-to-ground communication and multi-service carrying capacity.

2.3.3 Intelligent Operations and Maintenance Technology

• **Intelligent Passenger Flow Management:** Utilizes big data and AI algorithms for accurate passenger flow prediction, real-time monitoring, and intelligent guidance. Mature crowd management strategies have been developed for super transfer stations like Tiyu Xilu.

• **Intelligent Maintenance Platform:** Built an urban rail intelligent maintenance platform based on Digital Twin and Industrial Internet technologies, achieving real-time equipment status perception and Predictive Health Management (PHM).

- **Comprehensive Tunnel Inspection System:** Capable of performing sub-millimeter accuracy tunnel inspections at speeds of up to 60 km/h, with an efficiency 120 times greater than manual inspection, accurately detecting defects like cracks and misalignments.
- **360-Degree Train Inspection System:** Gradually replacing 80% of manual inspection tasks.
- **Smart Passenger Service:** Through the "Guangzhou Metro" App and smart stations, it provides convenient services such as the "One-Code Pass" for travel, intelligent travel guidance, and contactless payment, reshaping the passenger travel experience.

2.4 Pathways and Models of International Practice

Guangzhou Metro's internationalization represents the systematic export of its core capabilities and brand value, forming a multi-level, multi-path "Going Global" structure.

2.4.1 Model 1: High-End Consulting and Design Export ("Intellectual Export")

Case Study: Guangzhou Metro Design & Research Institute secured the overall design and general contract project for Ho Chi Minh City Metro Line 2 in Vietnam, marking its successful entry into the Vietnamese market and laying the foundation for expanding into the Southeast Asian market. Previously, it had already provided planning and consulting services for multiple countries in Asia and Africa.

2.4.2 Model 2: EPC Project Contracting ("Integrated Solution Export")

Case Study: Participating in international project bidding and construction through consortiums, responsible for the entire process from design and procurement to construction. For example, in collaboration with enterprises like China Railway Construction Corporation (CRCC), it successfully won the bid and constructed the Pakistan Lahore Orange Line Metro Train project. As a flagship project under the Belt and Road Initiative, it fully adopts Chinese standards, technology, and equipment.

2.4.3 Model 3: Operation Management and Training Services ("Management and Service Export")

Case Study: This is the most distinctive aspect of Guangzhou Metro's internationalization.

Pakistan Lahore Orange Line: Guangzhou Metro is responsible for the operation and maintenance services of this line, marking the first time its operational business has expanded abroad. It provided a full suite of services from operational preparation to long-term management and trained a significant number of local Pakistani technical personnel.

Domestic External Projects: Its operation management scope covers multiple external projects within China, including Nanchang Metro Line 3 (Jiangxi), Changsha Metro Line 6 (Hunan), Sanya Modern Tram (Hainan), Dongguan Huawei Songshan Lake Small Train, and Kunming Changshui International Airport Automated People Mover system.

2.4.4 Model 4: Technical Cooperation and Industrial Alliances ("Ecosystem Export")

Industrial Alliance Collaboration: Took the lead in establishing the Guangzhou Rail Transit Industry Alliance and founded the "Overseas Expansion Subcommittee" in 2025. This integrates industrial chain resources, adopts a "Collaborative Export + Integrated Export" model to address international competition, and enhances the overall international competitiveness of Guangzhou's rail transit industry.

International Technical Cooperation: In 2025, signed a Memorandum of Understanding with Singapore's SMRT Corporation. The collaboration focuses on sharing experiences and best practices in areas such as operation management, equipment maintenance, personnel training, smart technology R&D (e.g., AI, big data analytics), and commercial development.

The table below summarizes the main models and representative projects of Guangzhou Metro's international practice:

Table 2-2 Main Models and Representative Projects of Guangzhou Metro's International Practice

Model Type	Core Content	Representative Project/Initiative	Significance/ Impact
High-End Consulting & Design Export	Provides services such as network planning, feasibility studies, engineering design, and standards consulting	Overall Design and General Contract Project for Ho Chi Minh City Metro Line 2, Vietnam	Exports "top-level design" and "Chinese standards," representing the highest level of cooperation and paving the way for subsequent equipment and service exports.
EPC Project Contracting	Undertakes turnkey project contracting covering the entire process from design, procurement, to construction	Pakistan Lahore Orange Line Project (Belt and Road Flagship Project)	Comprehensively demonstrates the integration capabilities of Chinese enterprises and drives the holistic export of Chinese standards, technology, and equipment.
Operation Management & Training Services	Provides a full suite of services including operation preparation, operation management, maintenance, and personnel training.	Operation and Maintenance of Pakistan Lahore Orange Line; Operation of domestic projects like Nanchang Metro Line 3, Changsha Metro Line 6	Ensures the long-term success and sustainable development of projects, represents an in-depth export of technology and management experience, cultivates local talent, and enhances project sustainability.
Technical Cooperation	Takes the lead in forming	Establishment of	Integrates advantages of the

& Industrial Alliances	industrial alliances; engages in technical cooperation with international giants for joint development in third-party markets	Guangzhou Rail Transit Industry Alliance and its Overseas Expansion Subcommittee; Signing of MoU with Singapore's SMRT Corporation	industrial chain for "cluster-style" expansion overseas; achieves complementary advantages through international cooperation, learns advanced international experience, and enhances own competitiveness.
Brand & Technology Showcase	Participates in international high-end exhibitions to showcase advanced technological achievements and enhance international brand influence	Exhibition of the Tencent OS Intelligent Operating System at the China Pavilion, Expo 2025 Osaka, Kansai, Japan	Visually and vividly demonstrates the technical strength and innovative image of Chinese urban rail transit to the global public and business representatives, attracting potential partners.

2.5 In-depth Case Studies: Paradigms of International Practice

2.5.1 Pakistan Lahore Orange Line Project

This project is a quintessential example illustrating Guangzhou Metro's international practice.

- **Challenges: Political** and cultural differences, security risks, technical standards alignment, and insufficient skills among local staff.
- **Guangzhou Metro's Role: It** served not only as the operation service provider but also as a cultural bridge between China and Pakistan, an interpreter and adapter of technical standards, and the "chief instructor" for local employees.
- **Full Industrial Chain Export: The** project fully adopted Chinese standards, utilizing rolling stock and signaling systems adapted to the local environment. Guangzhou Metro localized its mature domestic operation management system and

emergency plans, ensuring the project's efficient and safe operation.

- **Vocational Education Practice: For** this project, Guangzhou Metro organized multiple large-scale technical and management training sessions in China for Pakistani trainees, covering all professional fields including dispatch, drivers, rolling stock, and power supply. This trained a large number of qualified local technical personnel, laying a solid foundation for the project's sustainable operation. This practice directly catalyzed and validated the feasibility and significant value of transforming the "Guangzhou Experience" into international vocational education.

2.5.2 Showcase at Expo 2025 Osaka, Japan

In 2025, Guangzhou Metro showcased its self-developed Suiteng OS intelligent operating system at the China Pavilion of Expo 2025 Osaka, Japan, vividly demonstrating to the world how smart technology is reshaping urban life.

- **Showcase Content: The** system demonstrated functions such as contactless fare gating, intelligent passenger flow management, one-click station startup/shutdown, video patrol inspection, and generating emergency dispatch plans within 15 seconds, highlighting AI's capability in resolving urban congestion.

- **"One-Code Pass" & "Four-Network Integration": Visitors** could experience cross-modal transfers using a "single code" for seamless travel across metro, regional railways, intercity railways, and trunk railways, offering an intuitive understanding of Guangzhou Metro's goals and achievements in advancing "Four-Network Integration."

- International Response: With approximately 12,000 visitors daily on average, it became one of the most attractive corporate brands in the China Pavilion, effectively enhancing Guangzhou Metro's international brand awareness and reputation and attracting potential international cooperation opportunities.

2.6 International Vocational Education: The Sublimation from Technical Practice to Knowledge Export

Through its international practice, Guangzhou Metro has profoundly realized that the export of technology and equipment is only the first step, while the export of talent and knowledge is the key to ensuring the long-term success, sustainable development, and local rooting of projects. This is the core theme of this book – International Vocational Education.

Formation of Vocational Education Content: Guangzhou Metro's international vocational education content originates from its core technologies, operating procedures, management norms, and emergency plans – encompassing both tacit and explicit knowledge – that have been repeatedly validated through practice over more than two decades in fields such as planning, design, construction, equipment, operation, maintenance, safety, and management. This knowledge has been systematically collated, summarized, refined, and standardized.

Advantages of the Vocational Education System:

(1) Practicality: Course content is derived from real-world complex scenarios and solutions, not purely theoretical teaching.

(2) Systematicness: Covers the entire lifecycle of urban rail transit, capable of providing complete training solutions from senior managers to front-line technical workers.

(3) Adaptability: Possesses the capability for localized customization according to the national conditions, standards, and cultures of different countries.

(4) Advancedness: Incorporates the latest achievements in smart rail transit and green rail transit technologies (e.g., Suiteng OS, intelligent maintenance).

Export Platforms and Models: Through various channels such as the Guangzhou Rail Transit Training College, industry-academia cooperation linked with the National Engineering Laboratory (e.g., joint postgraduate training with Hong Kong University of Science and Technology), undertaking national foreign aid training programs, and providing customized training for specific overseas projects (e.g., Lahore Orange Line), Guangzhou Metro is disseminating its knowledge system globally. Personnel training has also been explicitly identified as a core area in cooperation MoUs with international giants like Singapore's SMRT.

Chapter 3:International Export Pathways of Metro Technology under the Belt and Road Initiative

Against the backdrop of the steady advancement of the Belt and Road Initiative, China's rail transit industry has encountered historic opportunities for international development. As a leading enterprise in China's urban rail transit sector, Guangzhou Metro Group, leveraging its extensive construction and operation experience, advanced technical standards, and mature industrial chain system, has actively explored diversified and efficient international pathways for exporting metro technology. It provides high-quality rail transit solutions for partner countries involved in the Belt and Road Initiative. This chapter will systematically analyze Guangzhou Metro's model selection, key areas, standard internationalization strategy, and talent cultivation mechanisms during its technology export process, aiming to construct a comprehensive theoretical framework and practical pathway for the international export of rail transit technology.

3.1 The Belt and Road Initiative and Opportunities for Rail Transit Internationalization

Over the past decade since the implementation of the Belt and Road Initiative, infrastructure connectivity has consistently been a key area of cooperation. Rail transit, serving as the backbone of infrastructure, plays a vital role in this international collaboration. According to data from China's Ministry of Transport, as of the first half of 2024, China has engaged in railway cooperation projects with over 20 Belt and Road partner countries. These projects encompass full industrial chain cooperation including planning, design, construction, equipment manufacturing, operation, and maintenance, with cumulative contract values exceeding USD 50 billion.

The internationalization of China's rail transit is undergoing a transition from "hardware export" to "integrated hardware and software export." It is progressively shifting from singular engineering contracts to comprehensive export packages that include technical standards, operation management, and talent development. In May 2025, the "Rail Transit Standard Connectivity Scheme for Neighboring Countries" (RSCS Program), initiated by the China International Center for Sustainable Transport Innovation and Knowledge, was officially launched. This program aims to reduce technical barriers and enhance connectivity levels by coordinating technical standards and specifications, while also supporting neighboring countries in developing their own specific regulations and standards.

Guangzhou Metro Group has seized this historic opportunity to actively expand into international markets. Currently, enterprises within the Guangzhou rail transit industry chain have secured multiple overseas projects across 16 countries and regions in Asia, Europe, Africa, and South America. These projects cover various business types including design consultancy, construction, equipment manufacturing, and operation and maintenance, steadily moving towards providing global clients with one-stop solutions. Notably, the total rail transit mileage operated by Guangzhou Metro Group has reached 1,238 kilometers. This includes not only the local metro network but also intercity railway and overseas projects, such as the Pakistan Lahore Orange Line.

Table 3-1 International Business Expansion of Guangzhou Metro Group

Project Type	**Number of Countries Involved**	**Representative Project**	**Technology Export Content**
Design Consulting	8	Ho Chi Minh City Metro Line 2, Vietnam	Feasibility study, front-end engineering design
Construction	10	Lahore Orange Line, Pakistan	Full-system export of Chinese standards, technology, and equipment
Equipment Manufacturing	12	Bangkok Metro, Thailand	Core equipment including trains and signaling systems
Operation & Maintenance	6	Lagos Light Rail, Nigeria	Operation management, maintenance systems, talent

3.2 Models and Pathways for Guangzhou Metro's International Technology Export

Through years of international practice, Guangzhou Metro has gradually developed diversified and multi-level models and pathways for technology export, encompassing everything from simple equipment supply to the comprehensive export of complex technical standards and management systems. An analysis of Guangzhou Metro's international projects reveals the following primary technology export pathways:

3.2.1 Direct Technology Export Model

Direct technology export involves supplying hardware equipment such as metro vehicles, signaling systems, and automatic fare collection systems, along with technical services, directly to overseas markets through international trade and engineering contracts. This model represents the initial stage of Guangzhou Metro's internationalization but remains a significant form of export.

In September 2025, a consortium led by Guangzhou Metro Design & Research Institute Co., Ltd. signed a contract for the "Adjustment Feasibility Study Report Preparation, Front-End Engineering Design, and Procurement Services Project for the Ben Thanh-Tham Luong Section of Ho Chi Minh City Urban Railway Line 2, Vietnam," with a total contract value of approximately RMB 47.35 million. This project marks the first time Guangzhou Metro Design & Research Institute has directly secured an overseas overall general contract design project for rail transit, signifying its successful entry into the Vietnamese market. The successful signing of this project is not only effective proof that its brand influence and core competitiveness are recognized by the international market but also an important measure of its deep participation in the rail transit construction of Belt and Road partner countries.

3.2.2 Integrated Industrial Chain Export Model

The integrated industrial chain export model involves exporting the entire rail transit industrial chain—including planning and design, engineering construction, equipment manufacturing, and operation and maintenance—as a whole to overseas markets. This model can fully leverage the overall advantages of China's rail transit industry to achieve greater value creation.

The Pakistan Lahore Orange Line project is a paradigm of this model. As Pakistan's first metro and the first large-scale rail transit project launched and operational under the China-Pakistan Economic Corridor, it fully adopts Chinese standards, technology, and equipment. During construction, the project created over 7,000 local jobs. Since its official operation in 2020, the operational phase has created nearly 2,000 job opportunities, receiving high praise from local residents and becoming a new symbol of China-Pakistan friendship. The rigorous work ethic, superb professional skills, and high sense of responsibility of Guangzhou Metro staff have also motivated and inspired their Pakistani counterparts. Comprehensive training, from train operation and station services to equipment maintenance and emergency response, has enabled local employees to quickly master various skills required for metro operation. This has resulted in the Lahore Orange Line achieving on-time rates and service delivery rates both exceeding 99%, with all operational indicators maintained at a very high level.

3.2.3 Management and Standards Export Model

The management and standards export model involves the international transfer of soft power through operational management services, technical standards transfer, and talent cultivation. This model represents an advanced stage of rail transit technology export and can enhance the international influence and discourse power of China's rail transit industry.

The cooperation between Chongqing Transportation Investment Group and the Hanoi Metropolitan Railway Management Committee in Vietnam exemplifies this model. In June 2025, the two parties signed a Memorandum of Understanding in Chongqing. Chongqing Transportation Investment Group will share mature monorail technology and practical experience with Vietnam, contributing Chongqing's expertise to Hanoi's traffic governance. This cooperation is set against the backdrop of Hanoi entering a "fast track" of rail transit construction – with two lines planned to start construction in 2025, and several more rail transit lines expected to be completed by 2030 and 2045. Monorail lines are planned within this blueprint, creating an urgent need for the relevant technology.

3.2.4 International Cooperation and Joint Venture Model

The international cooperation and joint venture model involves collaborating with local or international enterprises to establish joint ventures or strategic alliances for jointly exploring international markets. This model can effectively reduce market entry barriers and improve project success rates.

In September 2025, Singapore's SMRT Corporation visited China's Guangzhou Metro Group and signed a Memorandum of Understanding. Both parties will share experiences and best practices in key areas such as operations management, equipment maintenance, personnel training, safety assurance, passenger experience, and environmental sustainability. Through mutual learning and technical exchanges, they aim to continuously enhance capabilities and professional expertise to better serve passengers. Simultaneously, the parties will also promote predictive maintenance and operational efficiency through joint research, development, and application of advanced technologies (including artificial intelligence, machine learning, and data analytics), and explore cooperation in commercial development within the rail transit sector.

Table 3-2 Comparison of Main Models for Guangzhou Metro's International Technology Export

Export Model	Main Characteristics	Applicable Conditions	Typical Case
Direct Technology Export	Primarily involves hardware equipment and technical services; simple and direct	Markets with significant technological gaps; initial entry stage	Ho Chi Minh City Metro Line 2 design project (Vietnam)
Integrated Industrial	Exports the entire	Markets of major	Lahore Orange Line

Chain Export	industrial chain; maximizes value creation	strategic importance; intergovernmental cooperation projects	project (Pakistan)
Management and Standards Export	Exports soft power; enhances international discourse power	Relatively mature markets; emphasis on long-term cooperative relationships	Chongqing-Hanoi monorail technology cooperation
International Cooperation & Joint Venture	Resource sharing, risk sharing, mutual benefit	Markets with high entry barriers requiring local knowledge	Guangzhou Metro - Singapore SMRT cooperation

3.3 Innovation and Export in Key Technology Areas

In its international technology export process, Guangzhou Metro focuses on integrating innovation achievements with practical applications, forming a number of key technology areas with international competitiveness. These technologies have not only been widely applied domestically but have also gained high recognition in the international market.

3.3.1 Smart Urban Rail Technology

Smart urban rail technology is one of the key areas for Guangzhou Metro's technology exports. With the rapid development of new-generation information technologies like artificial intelligence, big data, and cloud computing, Guangzhou Metro actively promotes the R&D and application of smart urban rail technologies and has successfully exported them to overseas markets.

At the 2025 China International Urban Rail Transit Exhibition, the smart urban rail solution jointly released by Jiadu Technology and Huawei, which deeply integrates AI and computing power, provided a new path for the digital transformation of urban rail. These innovative technologies include intelligent transport organization, intelligent passenger flow management, intelligent maintenance, and intelligent energy management systems, which can significantly improve metro operational efficiency and passenger experience.

The "AI+" action plan released by Qingdao Metro demonstrates the proactive initiatives of rail transit enterprises in embracing new technologies and leading change. These smart urban rail technologies are not only applied domestically but are also exported to overseas markets through Belt and Road cooperation. For instance, the intelligent dispatch system and intelligent ticketing system applied by Guangzhou Metro in the Pakistan Lahore Orange Line project have greatly enhanced operational efficiency and service levels.

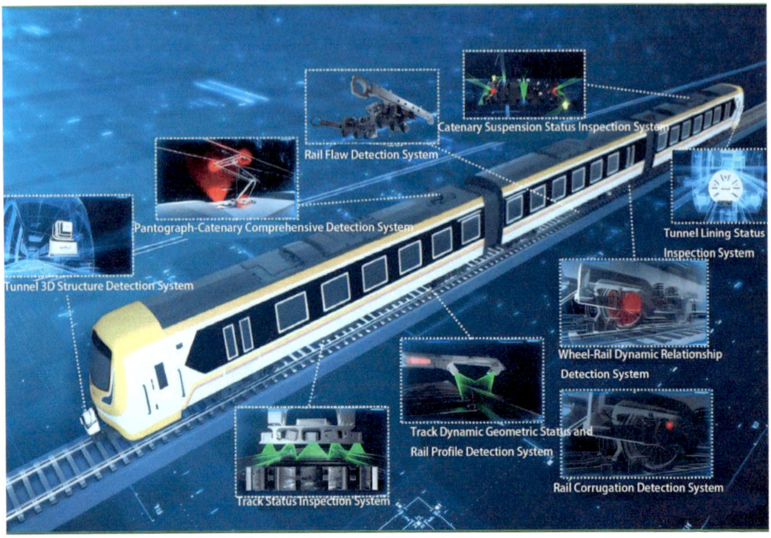

Figure 3-1 Smart Urban Rail Related Technologies

3.3.2 Straddle Monorail Technology

Straddle monorail technology represents another significant area of innovation in China's rail transit sector, offering distinct advantages particularly for addressing transportation challenges in mountainous and congested cities. Chongqing boasts the world's longest straddle monorail networkcarrying the highest daily passenger volume, with its rail transit system handling a remarkable 56.6% share of public transport trips.

Chongqing's core rail transit technologies encompass the entire chain including equipment manufacturing, system operation and maintenance, and talent cultivation. It has established a comprehensive straddle monorail standard system comprising over 60 standards and generated more than 200 technical patents. The entire industrial chain brings together over 150 entities, achieving an annual output value of approximately RMB 10 billion. In recent years, enterprises within Chongqing's monorail industrial chain have actively expanded globally, providing high-quality products and technical services for rail construction and operation in countries such as Thailand, Vietnam, Malaysia, Singapore, and Algeria. Newly signed contracts in 2024 alone exceeded RMB 4.2 billion.

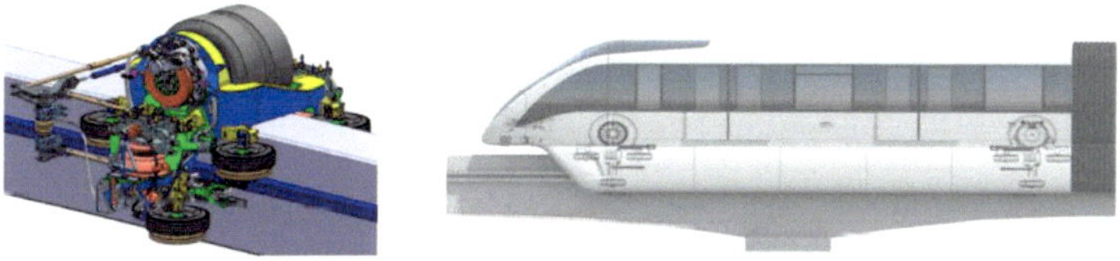

Figure 3-2 Straddle Monorail Diagram

3.3.3 Green and Low-Carbon Technologies

Green and low-carbon technologies represent another crucial area for Guangzhou Metro's technology exports. With increasing global emphasis on climate change and environmental protection, rail transit, as a green transportation mode, is seeing growing international demand for its low-carbon technologies.

Guangzhou Metro has undertaken significant innovation in green and low-carbon technologies, including regenerative braking energy recovery, intelligent lighting, and new energy applications. These technologies not only reduce the energy consumption and carbon emissions of metro operations but also minimize environmental impact. In international projects, Guangzhou Metro actively promotes these green and low-carbon technologies, assisting Belt and Road partner countries in achieving their sustainable development goals.

For example, in the Ho Chi Minh City Metro Line 2 project designed by Guangzhou Metro, green and low-carbon factors were fully considered. Energy-efficient trains, high-efficiency air conditioning systems, and intelligent energy management systems were adopted, which are projected to reduce the project's operational energy consumption by over 15%.

3.4 Standard Internationalization: The Core Strategy for Technology Export

Standard internationalization is the core strategy for the international export of

Chinese metro technology and a key link in upgrading from "technology export" to "standard export." For a long time, international rail transit standards were primarily dominated by developed countries in Europe and America, with China's rail transit industry having limited say in international standard setting. However, with the rapid development of China's rail transit technology and the acceleration of its internationalization process, this situation is gradually changing.

3.4.1 China's Efforts in Standard Internationalization

The Chinese rail transit industry places high importance on standard internationalization, promoting Chinese standards globally through various channels. In May 2025, the "Rail Transit Standard Connectivity Scheme for Neighboring Countries" (RSCS Program), initiated by the China International Center for Sustainable Transport Innovation and Knowledge, was officially launched. This program establishes a platform for rail transit standard cooperation among neighboring countries, aiming to reduce technical barriers and enhance connectivity levels by coordinating technical standards and specifications. The RSCS Program will serve as the foundation for rail transit standard connectivity among neighboring countries and support them in developing their own specific regulations and standards.

Regarding regional standard cooperation, the "China-ASEAN Railway Connectivity Standard Cooperation Alliance" was officially launched in June 2025. The alliance held its first plenary meeting in September 2025, adopting its charter and organizational structure, and establishing a council comprising representatives from government agencies responsible for railways and standardization, universities, research institutions, third-party organizations, and railway design units from China, Laos, Thailand, Cambodia, Myanmar, and Vietnam.

3.4.2 Guangzhou Metro's Practice in Standard Internationalization

Guangzhou Metro actively participates in standard internationalization work, promoting Chinese standards globally through project practice. In the Pakistan Lahore Orange Line project, Guangzhou Metro fully adopted Chinese standards, technology, and equipment, providing a successful case for the internationalization of Chinese standards. The successful implementation of this project demonstrates the advancement and applicability of Chinese standards, providing strong support for their promotion in the international market.

Furthermore, Guangzhou Metro actively participates in international standard setting. In September 2025, at the China Urban Rail Transit TOD Forum held in Guangzhou, Guangzhou Metro shared its standard-setting and practical experience in Transit-Oriented Development (TOD). These experiences are not only applied domestically but also provide reference for Belt and Road partner countries.

3.4.3 Challenges and Countermeasures for Standard Internationalization

Despite significant progress, the internationalization of Chinese rail transit standards still faces numerous challenges. Firstly, international standard competition is fierce; developed countries in Europe and America have long monopolized international standard setting, creating inertial resistance for Chinese standards "going global." Secondly, technical levels and demands vary greatly among countries, making standard coordination difficult. Thirdly, there is a lack of talent specialized in standard internationalization, hindering large-scale standard export.

To address these challenges, the Chinese rail transit industry can adopt the following countermeasures: First, strengthen cooperation with international standard organizations to increase China's voice in international standard setting. Second, focus on standard localization, adjusting standard content according to the actual conditions of different countries to improve adaptability. Third, strengthen the cultivation of talent for standard internationalization, building a professional team proficient in both technology and international rules.

3.5 Talent Cultivation and Vocational Education Cooperation: The Foundation for Sustainable Development

Talent cultivation and vocational education cooperation are vital components of international metro technology export and a crucial foundation for achieving sustainable development. During its internationalization process, Guangzhou Metro places high importance on talent cultivation and vocational education cooperation, training professional rail transit personnel for Belt and Road partner countries through various means.

3.5.1 International Talent Cultivation Models

Guangzhou Metro has explored and formed various international talent cultivation models, including training in China, local training, and joint cultivation programs. In 2024, Guangzhou Metro Group successfully held large-scale international exchange events such as the "China-ASEAN Rail Transit Sustainable Development Seminar," the "Advanced Study Program on Global Sustainable Transport and Cultural Exchange," and the "Ho Chi Minh City Urban Rail Transit 2024 (Guangzhou) Training Program," successfully converting market opportunities in Vietnam, Singapore, Malaysia, and others.

Regarding the Vietnamese market, multiple mutual visits at various levels from Guangzhou gained full recognition from all levels of the Vietnamese government, leading to the signing of cooperation intentions in talent training and TOD-related fields, continuously exporting Guangzhou Metro's construction, operation, and management concepts. The "Guangzhou Model" has become a benchmark for Ho Chi Minh City's future metro development.

3.5.2 Industry-Academia Cooperation for Talent Cultivation

Industry-academia cooperation is an important pathway for cultivating international rail transit talent. Southwest Jiaotong University, the National University of Laos, and the Railway Department of Laos' Ministry of Public Works and Transport signed a tripartite cooperation agreement, jointly establishing a "Joint Bachelor's-Master's-PhD Training" program, a "Short-term Training and Skills Enhancement" plan, and a "Joint Research and Standard Setting" platform. This industry-academia cooperation model leverages the strengths of all parties to cultivate high-level professional rail transit talent for Belt and Road partner countries.

3.5.3 Vocational Education Going Global

Along with the export of Chinese rail transit technology, vocational education is also going global. In its overseas projects, Guangzhou Metro focuses on cultivating local talent through methods like master-apprentice knowledge transfer, on-site training, and theoretical teaching to improve the skills of local employees. In the Pakistan Lahore Orange Line project, Guangzhou Metro trained a large number of local rail transit operation and management professionals, ensuring the project's sustainable development.

Despite notable achievements, the international export of Chinese metro technology still faces many challenges. Firstly, the international political and economic environment is complex and volatile, and geopolitical risks could affect the technology export process. Secondly, technical standards, laws, regulations, and cultural environments differ significantly among countries, posing adaptation challenges. Thirdly, international competition is intense, with traditional rail transit powerhouses like Japan and Germany still holding advantages in high-end markets. Fourthly, insufficient innovation in investment and financing models makes it difficult to meet diverse market demands.

To address these challenges, the Chinese rail transit industry can adopt the following countermeasures: First, strengthen country-specific research, deeply understand the market demands and investment environments of different countries, and formulate differentiated export strategies. Second, enhance technological innovation, improve the competitiveness of core technologies, especially in cutting-edge fields like smart urban rail and green low-carbon technologies. Third, innovate cooperation models, explore various models like BOT (Build-Operate-Transfer) and PPP (Public-Private Partnership) to reduce host countries' financing pressure. Fourth, strengthen risk prevention and control, establish a comprehensive risk assessment and prevention system to effectively respond to geopolitical risks.

With the deepening implementation of the Belt and Road Initiative and the continuous advancement of Chinese rail transit technology, the prospects for international export of Chinese metro technology are broad. Future trends for Chinese metro technology export include:

First, the export model will shift from "hardware export" to "integrated hardware and software export," placing greater emphasis on the export of soft power like standards, management, and operation. Second, the export fields will expand from traditional rail transit to emerging areas like smart urban rail and green low-carbon technologies. Third, the target markets will expand from developing countries to developed countries, achieving a comprehensive global market layout. Fourth, the export entities will shift from single enterprises to industrial chain clusters, forming advantages for integrated industrial chain export.

As a leading enterprise in China's rail transit industry, Guangzhou Metro will continue to actively explore new paths and models for the international export of metro technology under the Belt and Road Initiative, contributing Chinese wisdom and solutions to global rail transit development. Through deepening international cooperation, strengthening technological innovation, promoting standard internationalization, and emphasizing talent cultivation, Guangzhou Metro will achieve leapfrog development from "Going Out" to "Going In" and then to "Going Up," becoming a world-class comprehensive urban rail transit service provider.

In conclusion, the international export of metro technology under the Belt and Road Initiative is a systematic project requiring the joint efforts of multiple stakeholders including governments, enterprises, universities, and research institutions to build an open, inclusive, and mutually beneficial international cooperation ecosystem. Through the export of technology, standards, talent, and culture, the Chinese rail transit industry will provide high-quality rail transit solutions for Belt and Road partner countries, promoting sustainable regional economic and social development.

Part Ⅱ: Construction Technology

Chapter 4: Overview of Metro Engineering Construction

As the backbone of urban rail transit systems, the metro is an indispensable component of modern urban public transportation networks. After nearly three decades of development, Guangzhou Metro has progressed from constructing single lines to a new phase of large-scale, network-based development. It has established comprehensive capabilities encompassing the entire industrial chain, including planning and design, construction, equipment manufacturing, operation, and maintenance. This chapter provides a thorough overview of the development history, technological innovations, management practices, and international applications of Guangzhou Metro's engineering construction, offering readers a systematic and comprehensive perspective on metro engineering construction.

4.1 General Overview of Guangzhou Metro Engineering Construction

Since the opening of its first line in 1997, Guangzhou Metro has undergone rapid development, evolving from nothing to a fully-fledged system and from a single line to networked operations. As of June 2025, Guangzhou Metro has opened 19 lines with a total operational mileage of 751.1 kilometers, forming an extensive and efficient network. This scale not only ranks among the top in China but also makes Guangzhou one of the cities with the most developed metro networks in the world.

The construction philosophy of Guangzhou Metro has transitioned from "meeting basic travel needs" to "providing a high-quality travel experience." Early construction primarily addressed urban traffic congestion, while current efforts place greater emphasis on passenger experience, cultural heritage, and environmental protection. In the most recently opened lines, station designs fully integrate local cultural elements of Guangzhou. For instance, the second phase of Line 7 adopts the theme of "Blue-Green Landscapes," while the eastern extension of Line 5 employs the "One Station, One Color" design concept. These designs not only meet functional transportation requirements but also serve as windows showcasing the city's culture.

The construction management model of Guangzhou Metro has also continuously innovated and improved. From initially learning from the experience of Hong Kong MTR, it has gradually developed its own construction management standards and technical systems with Guangzhou characteristics. Guangzhou Metro now possesses full-process, whole-industrial-chain project management capabilities. Particularly in large-scale complex projects, such as the Line 11 loop line project, Guangzhou Metro has overcome numerous world-class technical challenges and achieved nearly 12 domestic or Guangzhou-first records.

Table 4-1 Guangzhou Metro Construction Milestones

Year	Construction Milestone	Significance
1997	Line 1 commenced operation	Guangzhou's first metro line, with a total length of 18.5 kilometers
2003	Line 2 commenced operation	First adoption of a domestically produced signaling system
2010	Zhujiang New Town Automated People Mover system commenced operation	First fully automated driverless line
2018	Achieved "metro access to all districts"	All 11 districts of the city are covered by metro service
2023	Operational mileage reached 653 kilometers	Became a world-class network
2025	Operational mileage reached 751.1 kilometers	Formed a "Loop + Cross + Radial" network structure

The achievements of Guangzhou Metro's construction are reflected not only in route length and transport capacity but also in technological innovation and standard development. Through continuous technological research, development, and practical application, Guangzhou Metro has established a comprehensive standardized system for engineering construction, covering all stages including design, construction, equipment, and operation. These standards not only ensure the high-quality construction of Guangzhou Metro but have also laid a solid foundation for the standardization and internationalization of urban rail transit in China.

4.2 Core Technologies and Innovations in Metro Engineering Construction

Metro engineering construction is a complex systematic project involving multiple stages such as civil construction, equipment installation, and commissioning operations, requiring the comprehensive application of various technologies and technological innovation. Through long-term construction practice, Guangzhou Metro has developed a series of core technologies and innovative achievements.

4.2.1 Innovations in Shield Tunneling Technology

Shield tunneling technology is a critical component of metro tunnel construction. Guangzhou's complex geological conditions, often referred to as a "geological museum," include soft soil, sand layers, hard rock, and mixed strata, presenting extremely high demands for shield tunneling. To address these complex geological conditions, Guangzhou Metro builders have developed and applied multiple innovative technologies.

During the construction of Line 7 Phase 2, the engineering team successfully deployed the nation's first triple-mode shield machine. This machine can freely switch between Earth Pressure Balance (EPB) mode, Slurry Mode, and Open Mode based on changes in geological conditions, significantly enhancing construction efficiency and safety. In the construction of Line 11, the Liuhua Station utilized the Pile-Beam-Arch (PBA) method, becoming the first station in South China to employ the PBA method in water-rich rock strata.

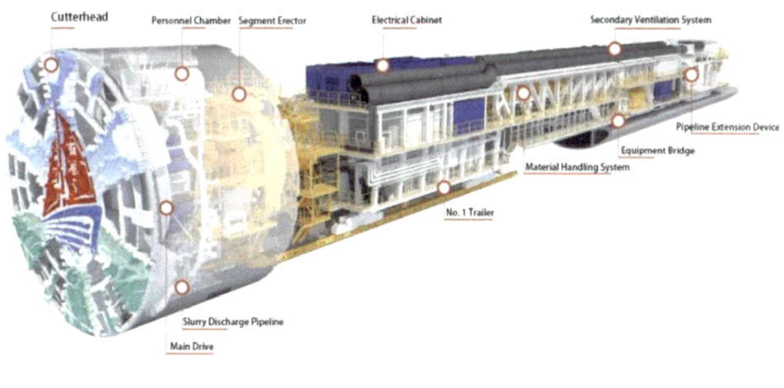

Figure 4-1 The Nation's First Triple-Mode Shield Machine

Facing the boulder-rich ground conditions, the construction team employed a combination of detection methods including surface micro-tremor survey and geological drilling. They implemented pre-blasting treatment and grouting reinforcement in advance, effectively resolving major obstacles during construction. These technological innovations not only ensured the smooth progress of the project but also accumulated valuable experience for metro construction under similar geological conditions.

4.2.2 Intelligent Construction Technology

With the advancement of information technology, intelligent construction has become a crucial direction in metro development. Guangzhou Metro actively promotes the application of new technologies such as BIM, IoT, and big data in engineering construction, significantly enhancing both quality and efficiency.

During the construction of Guangzhou Metro Line 11, the nation's first building component assembly robot, named "Chisha," was introduced. This robot can navigate uneven stairs and is equipped with keen "eyes." It requires only one operator and one director, greatly reducing the labor intensity of construction personnel. Furthermore, its work efficiency has increased by over 1.5 times, while simultaneously significantly reducing construction noise and dust generation. The development of the "Chisha" robot fills a gap in the field of prefabricated building construction equipment in China.

Guangzhou Metro has also extensively deployed a track area dispatch management system, video surveillance system, and facial recognition system, achieving comprehensive monitoring coverage of personnel and equipment across both time and space. This enables visualized, intelligent, and controllable safety management. During the construction of Line 21, multiple rounds of optimization were conducted on construction plans through information technology and smart construction platforms, providing sufficient lead time for problem resolution.

4.2.3 Vibration and Noise Reduction Technology

Vibration and noise generated by metro operations are significant urban environmental concerns. Guangzhou Metro places great emphasis on the application of vibration and noise reduction technologies during engineering construction to minimize the impact on the surrounding environment.

During the construction of Guangzhou Metro Line 21, the initial environmental impact assessment identified 88 sensitive points along the entire line. To better mitigate impacts, Guangzhou Metro organized a secondary expert assessment, increasing the number of identified sensitive points to over 100. Throughout the track engineering construction, builders installed a total of 48.69 kilometers of various vibration-damping track beds at these environmentally sensitive locations. This included 14.45 kilometers of medium-damping track beds, 21.19 kilometers of high-damping track beds, and 13.05 kilometers of special-damping track beds, effectively reducing the impact of metro operations on the surrounding environment.

Particularly in densely built-up areas such as Yuancun, Tianhe Smart City, Science City, and Zhenlongxu, the internationally advanced liquid-damped steel spring floating slab technology – offering superior vibration damping performance – was adopted. The expected vibration reduction effect exceeds 15 decibels. Simultaneously, integrated measures combining track vibration reduction with sound barriers were implemented in elevated sections, minimizing the environmental impact of both vibration and noise.

4.2.4 Safety and Quality Control Technology

Safety and quality are the lifelines of metro engineering. Through long-term practice, Guangzhou Metro has developed a comprehensive safety and quality control system alongside technical measures.

During track construction, a "sample section first" approach was adopted. Before commencing each construction process, a sample section was inspected and accepted. Work proceeded fully only after meeting standards. Field quality control responsibilities were strictly enforced, with audio-visual records maintained for inspections of each process, ensuring clear accountability and quality assurance at every post. All 13 kilometers of special-damping track bed (liquid-damped steel spring floating slab) along the entire line utilized a prefabricated slab structure. This not only improved the concrete quality of the track bed, facilitating future maintenance, but also increased track-laying progress by 20% compared to cast-in-place floating slab track beds.

To address the challenge of large amounts of construction wastewater generated during the construction of support structures and earth excavation at metro stations, the builders developed a hydraulic filter press system that is effective, efficient, low-cost, and easy to operate. This system can classify and treat wastewater and solid waste. The wastewater, after treatment, becomes clear water that meets all standards for discharge into the municipal (pipe network) and can also be reused for dust suppression spraying, green plant maintenance, and site cleaning, achieving "zero discharge" of construction silt-laden water pollutants.

Table 4-2 Major Technological Innovations in Guangzhou Metro Engineering Construction

Technology Area	Innovation Content	Application Effectiveness
Shield Tunneling Technology	Triple-mode shield machine, PBA (Pile-Beam-Arch) construction method	Adapts to complex geological conditions and improves construction efficiency
Intelligent Construction	Building component assembly robot, smart construction platform	Increases work efficiency by 1.5 times and reduces noise and dust
Vibration and Noise Reduction	Liquid-damped steel spring floating slab, integrated vibration damping measures	Achieves vibration reduction effect exceeding 15 decibels
Safety and Quality	"Sample section first" approach, hydraulic filter press system	Achieves "zero discharge" of construction silt-laden water pollutants

4.3 International Application and Export of Metro Engineering Construction

With the deepening of the Belt and Road Initiative, Guangzhou Metro, leveraging its extensive construction experience and technological accumulation, has actively explored international markets, achieving a strategic shift from "bringing in" to "going global." Guangzhou Metro's engineering technologies and management experience have been successfully exported to multiple countries, providing Chinese solutions and wisdom for the development of urban rail transit worldwide.

4.3.1 Breakthrough in the Vietnamese Market

In September 2025, a consortium led by Guangzhou Metro Design & Research Institute Co., Ltd. signed a contract for the "Adjustment Feasibility Study Report Preparation, Front-End Engineering Design, and Procurement Services Project for the Ben Thanh-Tham Luong Section of Ho Chi Minh City Urban Railway Line 2, Vietnam," with a total contract value of approximately RMB 47.35 million. This project marks the first time Guangzhou Metro Design & Research Institute has directly secured an overseas overall general contract design project for rail transit, signifying its successful entry into the Vietnamese market.

The successful signing of this project is not only effective proof that its brand influence and core competitiveness are recognized by the international market but also an important measure of its deep participation in the rail transit construction of Belt and Road partner countries. Ho Chi Minh City is Vietnam's largest port city and economic center, and a city for international friendly cooperation and exchange with Guangzhou. Currently, Ho Chi Minh City is significantly accelerating its urban rail construction, planning to complete a new urban railway network by 2035.

In 2023, the China Association of Metros, Guangzhou Metro Group, and the Ho Chi Minh City Management Authority for Urban Railway (MAUR) signed a Memorandum of Understanding in Guangzhou. They agreed to work together to set common goals supporting the research and implementation of the development strategy for Ho Chi Minh City's urban railway system. The tripartite cooperation specifically includes sharing China's experience and solutions in preliminary rail transit planning, cooperation in knowledge sharing and talent cultivation, and collaboration in sharing construction techniques.

4.3.2 Project Practice in Pakistan

In 2020, a consortium comprising Guangzhou Metro Group, Norinco

International, and Pakistan's DW Company signed an operation and maintenance service contract for the Lahore Orange Line Metro Train project with the Punjab Mass Transit Authority of Pakistan. Guangzhou Metro operating and maintaining Pakistan's first metro line marked the first time its operation and maintenance business expanded abroad, also highlighting the international influence of Guangzhou's rail transit industry.

The Lahore Orange Line project in Pakistan is the first large-scale rail transit project under the China-Pakistan Economic Corridor of the Belt and Road Initiative, fully adopting Chinese standards, technology, and equipment. During its construction phase, the project created over 7,000 local jobs, and the operational phase has created nearly 2,000 employment opportunities, receiving high praise from local residents and becoming a new symbol of China-Pakistan friendship.

4.3.3 Technical Cooperation with Singapore

In October 2024, Guangzhou Metro and Singapore Mass Rapid Transit (SMRT) Trains Ltd signed a Memorandum of Understanding (MoU). Both parties expressed their desire to establish a long-term strategic partnership, leveraging their respective strengths and expertise to advance rail technology, operations, and maintenance efficiency in both cities.

As two internationally leading metro operators, Guangzhou Metro and SMRT had already engaged in in-depth exchanges and reached cooperation intentions in operational technologies such as tunnel structure inspection and track renewal, as well as in commercial areas like cross-border advertising agency and TOD business. Within the signed MoU, both parties explicitly agreed to strengthen cooperation and exchange focusing on rail transit operation and maintenance, particularly in metro station energy management, metro tunnel inspection and monitoring, track facility renewal, and full lifecycle system management of rail transit based on safety, reliability, and intelligence.

4.3.4 International Market Expansion Strategy

In its international market expansion, Guangzhou Metro has adopted a diversified and comprehensive strategy. On one hand, it achieves value chain upgrading through the export of technology, management, and standards; on the other hand, it enhances technological innovation capability through industry-academia-research collaboration to support internationalization.

In 2024, Guangzhou Metro Group promoted the comprehensive "going global" of its industrial chain, successfully organizing large international exchange events such as the "China-ASEAN Rail Transit Sustainable Development Seminar," the "Advanced Study Program on Global Sustainable Transport and Cultural Exchange," and the "Ho Chi Minh City Urban Rail Transit 2024 (Guangzhou) Training Program," successfully converting market opportunities in Vietnam, Singapore, Malaysia, and others.

Regarding the Vietnamese market, multiple mutual visits at various levels from Guangzhou gained full recognition from all levels of the Vietnamese government, leading to the signing of cooperation intentions in talent training and TOD-related fields, and continuously exporting Guangzhou Metro's construction, operation, and management concepts. The "Guangzhou Model" has become a benchmark for Ho Chi Minh City's future metro development.

Furthermore, Guangzhou Metro also secured the operation and maintenance service project for the first phase of the Abuja Light Rail in Nigeria, successfully exploring the African market and achieving the first export of its intercity operation business; the Lahore project in Pakistan achieved profit repatriation for the first time; and it reached strategic cooperation agreements with the Ho Chi Minh City Management Authority for Urban Railway (MAUR), Vietnam's CT Group, Singapore's SMRT Trains, and Istanbul Metro of Turkey, among others, exploring and simultaneously advancing rail transit markets in countries such as Vietnam, Cambodia, Colombia, São Paulo (Brazil), and Serbia.

Table 4-3 Guangzhou Metro International Business Expansion

Country/Region	**Project Type**	**Technology Export Content**	**Characteristics & Significance**
Vietnam	Design Consulting	Feasibility study, front-end	First overseas overall design

		engineering design	contract project
Pakistan	Operation & Maintenance	Full-system export of Chinese standards, technology, and equipment	First overseas operation and maintenance business
Singapore	Technical Cooperation	Tunnel inspection, track renewal, energy management	Strategic cooperation with internationally leading metro company
Nigeria	Operation & Maintenance	Intercity railway operation and maintenance	Successful entry into African market

4.4 Talent Cultivation and Vocational Education in Metro Engineering Construction

Metro engineering construction requires a large number of professional technical personnel and highly skilled talents. Talent cultivation and vocational education are crucial foundations supporting the sustainable development of metro construction. Throughout its long-term development, Guangzhou Metro has formed a comprehensive talent cultivation system and vocational education model.

4.4.1 Talent Cultivation Model Integrating Industry, Academia, and Research

Guangzhou Metro highly values the integration of industry, academia, and research, cultivating high-quality professional technical talents through various means such as cooperation with universities, establishing R&D institutions, and engaging in international collaboration.

Regarding R&D platform development, the Guangzhou Metro National Engineering Laboratory has established a joint postgraduate training cooperation mechanism with the Hong Kong University of Science and Technology. They jointly cultivate PhD students in three key areas: demand forecasting and planning for metropolitan area transportation systems based on big data, AI-powered rail transit operation management, and integrated transportation information sharing with multi-modal coupling. This provides strong talent and intellectual support for the development of Guangzhou's rail transit industry.

In standard setting and skills certification, Guangzhou Metro actively participates in industry standard development. Through summarizing practical experience, it has formed a complete standard system for engineering construction. These standards are not only applied in Guangzhou Metro's construction practice but also serve as important content for vocational education and training, providing a normative basis for industry talent cultivation.

4.4.2 International Talent Cultivation

With the acceleration of internationalization, Guangzhou Metro has intensified its efforts in cultivating international talents. Through international project practice, overseas training, and international cooperation exchanges, it has cultivated a group of interdisciplinary talents proficient in both technology and international rules.

Large international exchange events successfully organized by Guangzhou Metro, such as the "China-ASEAN Rail Transit Sustainable Development Seminar," the "Advanced Study Program on Global Sustainable Transport and Cultural Exchange," and the "Ho Chi Minh City Urban Rail Transit 2024 (Guangzhou) Training Program," not only promote Guangzhou Metro's technology and experience but also provide a platform for cultivating rail transit talents domestically and internationally.

In international project practice, Guangzhou Metro focuses on localized talent cultivation, improving the skill levels of local employees through methods like master-apprentice knowledge transfer, on-site training, and theoretical teaching. In the Pakistan Lahore Orange Line project, Guangzhou Metro trained a large number of local rail transit operation and management professionals, ensuring the project's sustainable development.

4.4.3 International Cooperation in Vocational Education

Guangzhou Metro actively engages in international cooperation in vocational education, enhancing its internationalization level through introducing international certifications, cooperative education programs, and faculty exchanges.

The cooperation between Guangzhou Metro and Singapore MRT is not limited to the technical field but also includes talent cultivation. The MoU clearly defines cooperation content in knowledge sharing and talent cultivation, which will provide new opportunities for Guangzhou Metro to cultivate international talents.

Furthermore, by participating in international conferences, exhibitions, and technical exchange activities, Guangzhou Metro continuously learns and introduces advanced international experience and technology, while simultaneously showcasing its achievements and experience to the world, enhancing its international influence and discourse power.

4.5 Future Development Trends and Challenges

With technological progress and evolving urban development needs, metro engineering construction faces new development trends and challenges. Guangzhou Metro needs to accurately grasp these trends, address the challenges, and maintain sustainable development.

4.5.1 Smart Metro Construction

Smart metro is a crucial direction for future metro development. By applying new technologies such as artificial intelligence, big data, and the Internet of Things, it aims to achieve intelligence in metro planning, design, construction, operation, and service, enhancing the safety, efficiency, and passenger experience of the metro system.

Guangzhou Metro has already actively explored smart metro construction. For instance, the smart urban rail solution jointly released by PATEO and Huawei deeply integrates AI and computing power, providing a new path for the digital transformation of urban rail. These innovative technologies include intelligent transport organization, intelligent passenger flow management, intelligent maintenance, and intelligent energy management systems, which can significantly improve metro operational efficiency and passenger experience.

In the future, Guangzhou Metro needs to further strengthen the R&D and application of smart metro technologies, especially in areas of self-sufficient and controllable technologies, such as signaling systems and intelligent maintenance systems with fully independent intellectual property rights, to enhance technical security and competitiveness.

4.5.2 Green and Low-Carbon Development

Green and low-carbon development is a significant trend in metro construction. By applying energy-saving technologies, green materials, and low-carbon processes, it aims to reduce the energy consumption and emissions of metro construction and operation, achieving sustainable development.

Guangzhou Metro has undertaken significant innovation in green and low-carbon technologies, including regenerative braking energy recovery, intelligent lighting, and new energy applications. These technologies can reduce the energy consumption and carbon emissions of metro operations and minimize environmental impact.

In future construction, Guangzhou Metro needs to further strengthen the green concept, implementing green and low-carbon requirements throughout the entire process from planning and design, material selection, and construction techniques to operation management, building a resource-saving and environmentally friendly metro system.

4.5.3 Challenges of Super-Large-Scale Network Operation

With the continuous expansion of the network scale, Guangzhou Metro faces the challenges of super-large-scale network operation. These include issues related to network operation management, aging equipment governance, passenger flow organization optimization, and emergency response capability enhancement.

To address these problems, Guangzhou Metro needs to strengthen research on network operation theory and technology, improve operation management standards and norms, and enhance network coordination efficiency and reliability. Simultaneously, it needs to strengthen the health management and preventive maintenance of equipment and facilities to ensure they remain in good condition.

4.5.4 Challenges in International Development

In the process of internationalization, Guangzhou Metro faces multiple challenges such as differences in international standards, cultural differences, and political risks. Technical standards, laws, regulations, and cultural environments vary significantly across countries, requiring targeted strategies to address them.

To cope with these challenges, Guangzhou Metro needs to strengthen country-specific research, deeply understand the market demands and investment environments of different countries, and formulate differentiated export strategies. Concurrently, it needs to strengthen risk prevention and control, establishing a comprehensive risk assessment and prevention system to effectively respond to geopolitical risks.

Furthermore, Guangzhou Metro needs to strengthen the cultivation of international talents, building a professional team proficient in both technology and international rules, to provide talent support for international development.

Metro engineering construction is a complex systematic project involving the collaboration of multiple disciplines and stages. After nearly three decades of development, Guangzhou Metro has formed a complete technical system and management standards, accumulating rich construction experience. Through continuous technological and management innovation, Guangzhou Metro has successfully built a world-class urban rail transit network, providing strong support for urban development and the construction of the Greater Bay Area.

With the deepening of the Belt and Road Initiative, Guangzhou Metro has actively explored international markets, achieving a strategic shift from "bringing in" to "going global." Through the export of technology, management, and standards, Guangzhou Metro has provided Chinese solutions and wisdom for the development of urban rail transit worldwide, promoting the internationalization of Chinese urban rail transit standards.

In the future, Guangzhou Metro needs to grasp development trends such as smart metro and green low-carbon development, address the challenges of super-large-scale network operation and international development, maintain its innovative vitality and development momentum, and make new contributions to urban rail transit development.

Chapter 5:Tunnel Construction and Shield Tunneling Technology

Tunnel construction represents the core component of metro development, whose technical proficiency directly impacts project quality, safety, and progress. Through nearly three decades of development, Guangzhou Metro has achieved globally remarkable accomplishments in the field of tunnel construction and shield tunneling technology. It has not only overcome a series of technical challenges presented by the "geological museum" but has also established a technologically innovative system and standards with international influence. This chapter will systematically elaborate on Guangzhou Metro's innovative practices, technological achievements in tunnel construction and shield tunneling technology, and their application and export in the field of international vocational education.

5.1 Geological Characteristics of Guangzhou and Tunnel Engineering Challenges

The geological conditions in the Guangzhou area are extremely complex, earning it the reputation as a "geological museum" as it encompasses almost all typical geological scenarios. Ranging from soft soil, sand layers, and gravelly sand to hard granite, and from massive karst caves to hazardous fault zones, these geological conditions pose severe challenges for tunnel construction.

The primary geological challenges faced by Guangzhou Metro tunnel construction include: the high risks in karst development areas, the instability of strata with "upper-soft and lower-hard" composition, the flow characteristics of high-water-table, water-rich sand layers, and the potential hazards of fault zones. For instance, in the Jiangshui section of Guangzhou Metro Line 8's northern extension, the groundwater level in the left tunnel area is high. The section passes through strata including silty clay, gravelly sand, silty fine sand, and medium-coarse sand, characterized by widely distributed water-rich sand layers, loose geological structure, and strata with high permeability coefficients. Simultaneously, karst development in this area is significant, with a cavity encounter rate reaching 45.1%, and the distribution is irregular. Similarly, in the Zhongjiang section of the branch line of Line 8's northern extension, limestone accounts for 58% of the full face, the cavity encounter rate within the section is as high as 57.8%, and the shield receiving section is located in a full-face sand layer geological condition.

This complex geological environment presents multiple challenges for shield tunneling: during shield advancement, the cutterhead center area is prone to forming a "mud cake"; preventing ground settlement and machine subsidence becomes a critical and difficult aspect of construction; in karst development areas with high cavity encounter rates, the shield machine faces risks of sinking and nose-diving; and single-mode shield machines have limitations when constructing in areas with "upper-soft and lower-hard" strata and significant karst development.

Table 5-1 Major Geological Challenges and Solutions in Guangzhou Metro Tunnel Engineering

Geological Challenge	Typical Characteristics	Construction Risks	Innovative Solutions
Karst Development Area	Cavity encounter rate 45.1%-57.8%	Shield machine sinking, nose-diving risk	Advanced geological prediction, karst cave pre-treatment
Water-Rich Sand Layer	High groundwater level, high stratum permeability coefficient	Ground settlement, machine subsidence	Slurry balance shield, synchronous two-component grouting
Upper-Soft Lower-Hard Stratum	Alternating soft soil and hard rock	Cutterhead wear, difficult attitude control	Triple-mode shield machine, dynamic mode switching
High-Hardness Rock Layer	Rock strength up to 147 MPa	Low tunneling efficiency, rapid tool wear	TBM mode, roller cutter technology

5.2 Guangzhou's Contributions to Shield Tunneling Technology Innovation

Confronted with complex geological challenges, Guangzhou Metro collaborated with design institutes, construction companies, scientific research institutions, and universities to conduct a series of technological breakthroughs and innovations, forming a shield tunneling technology system at an internationally leading level.

5.2.1 "Hengdun Mud" Technology: An Internationally Leading Innovative Breakthrough

In 2017, the research project "Key Technology Research on 'Hengdun Mud' Assisted Pressure Chamber Entry in Shield Tunneling Construction," led by Guangzhou Metro Group Co., Ltd., passed the scientific and technological achievement appraisal and was recognized as "filling an international technological gap and fully reaching the international leading level." Academician Qian Qihu of the Chinese Academy of Engineering highly praised the technology, stating: "'Hengdun Mud' technology reduces the construction risks associated with pressure chamber entry during shield tunneling, greatly enhances China's shield tunneling engineering technical level, and holds milestone significance for shield tunneling construction technology."

"Hengdun Mud" technology is a significant innovative achievement within Guangzhou Metro's "Theoretical System for Shield Tunneling in Composite Strata." This technology integrates methods, materials, and equipment into a systematic technique, overcoming technical difficulties such as high risk, high investment, and longconstruction period for shield chamber opening operations. Compared with traditional techniques, "Hengdun Mud" technology offers advantages like lower cost (single application cost approximately RMB 300,000, significantly lower than traditional reinforcement methods costing millions), long stabilization time, and strong adaptability. Particularly under special conditions such as beneath rivers or structures, it enables efficient and safe shield chamber opening operations.

To date, "Hengdun Mud" technology has been successfully applied in 9 cities, across 27 projects, and on 56 shield machines. It has successfully withstood tests in special high-risk projects such as Fuzhou Metro shields crossing the Min River, Xiamen Metro shields crossing the sea, and Lanzhou Metro shields crossing the Yellow River.

5.2.2 Triple-Mode Shield Machine: A Technologically Pioneering Leap

In 2023, the scientific research project "Key Technology and Application of Triple-Mode Tunnel Boring Machine for Urban Tunnels in Ultra-Hard Rock Composite Strata," undertaken by Guangzhou Metro Group, passed the scientific and technological achievement appraisal. The appraisal committee, chaired by Academicians Qian Qihu and Chen Xiangsheng of the Chinese Academy of Engineering, and Chief Engineer Xu Tianping of the Provincial Construction Group, unanimously concluded that the project reached the "international leading level."

The triple-mode shield machine is the world's first tunneling equipment integrating three modes: slurry, earth pressure balance (EPB), and Tunnel Boring Machine (TBM). It enables efficient tunneling, safety management, and precise control in urban tunnels within ultra-hard rock composite strata. This equipment introduces the concept of balanced tunneling with rapid establishment of earth chamber pressure into the basic TBM function of crushing ultra-hard rock, breaking through the industry challenge of balanced continuous tunneling with dual-channel muck discharge in water-rich fault zones within hard rock. The series of technologies for multi-mode conversion in the hard rock TBM are an international innovation.

This innovative technology has been successfully applied in the Luogang to Shuixi section of Guangzhou Metro Line 7 Phase 2. In this section, the maximum strength of the full-face hard rock segment reaches 147 MPa. The average daily tunneling speed of the shield machine was 2.62 times that of EPB shields and 1.45 times that of EPB/TBM dual-mode shields in similar strata within the same area, achieving significant economic and social benefits. Notably, this technology also won the International Tunneling and Underground Space Association (ITA) Annual Product/Equipment Innovation Award in 2023, recognized as the "Oscar Award" of the international tunneling industry.

5.2.3 Dual-Mode Shield Machine: A Flexible Solution for Complex Geology

In addition to the triple-mode shield machine, Guangzhou Metro also developed and applied EPB/Slurry dual-mode shield machines, providing more flexible solutions for different geological conditions. In the branch line project of Line 8's northern extension, constructors faced the complex geological conditions of the Zhongjiang section (including karst caves, fault zones, and upper-soft lower-hard strata). After multiple rounds of thematic discussions and field investigations, it was ultimately decided to change the original slurry shield machine to an EPB/Slurry dual-mode shield machine.

This dual-mode shield machine has a diameter of 6.69 meters, a total length of about 110 meters, a main machine length of about 10 meters (including the cutterhead), and a total weight of about 635 tons. It can flexibly switch between Earth Pressure Balance mode and Slurry Mode according to changes in geological conditions, greatly improving the adaptability and safety of the shield machine under varying geological conditions.

Table 5-2 Comparison of Major Shield Tunneling Technology Innovations at Guangzhou Metro

Technology Type	Technical Characteristics	Application Scenario	Technical Advantages	Application Effectiveness
Hengdun Mud Technology	Integrates method, material, and equipment	Shield chamber opening operations; special working conditions	Low cost, high safety, long stabilization time	Applied in 9 cities, 27 projects, across 56 shield machines
Triple-Mode Shield Machine	Slurry + EPB + TBM three modes	Ultra-hard rock composite strata	Rapid mode switching without disassembly/assembly	Tunneling speed increased by 2.62 times
Dual-Mode Shield Machine	EPB + Slurry two modes	Karst development areas; upper-soft lower-hard strata	Improved adaptability and safety	Handled cavity encounter rate of 57.8%
Slurry Balance Shield	Air cushion direct discharge dual-channel	High-water-table, water-rich sand layers	Prevents cutterhead mud cake formation; controls settlement	Achieved geological ahead of cutterhead up to 30 meters

5.3 Key Technologies and Refined Management in Tunnel Construction

In tunnel construction, Guangzhou Metro not only focuses on technological innovation but also implements refined management to ensure project safety, quality, and progress. The following are detailed introductions to several key technologies:

5.3.1 Advanced Geological Prediction and Karst Cave Treatment Technology

To address the high risks in karst development areas, Guangzhou Metro has developed a complete set of advanced geological prediction and karst cave treatment technologies. During the construction of Line 8's northern extension, builders adopted an integrated approach of "geophysical prospecting + drilling technology" and the "dual-fluid sealing edge + grouting filling process" to pre-lay an "artificially reinforced channel" for the shield machine.

Before construction, detailed surveys and assessments were conducted. The shield machine was equipped with an advanced geological prediction system. By integrating seismic wave prediction results, geological survey data, tunneling parameters, and surface monitoring data, comprehensive forecasting and mutual verification were performed. This achieved geological imaging up to 30 meters ahead of the cutterhead, enabling real-time monitoring of geological changes and dynamic adjustment of tunneling parameters to ensure safe passage.

For areas with strong karst development (such as the Zhongjiang section with a cavity encounter rate of 57.8%), the project team used methods like drilling coring and ground-penetrating radar to accurately detect the distribution of karst caves and fissures, and the properties of fillers. Pre-treatment grouting was performed on karst caves within the tunnel range to ensure stratum stability and prevent the shield machine from "nose-diving" or deviating sideways.

5.3.2 Customized Design and Application of Shield Machines

Guangzhou Metro adheres to the "one section, one strategy" design philosophy, customizing shield machines according to the specific geological conditions of each tunnel section. In the Jiangshui section of Line 8's northern extension, characterized by high groundwater levels and widespread water-rich sand layers, Guangzhou Metro collaborated with PowerChina and Guangzhou Metro Design Institute 等单位 to tackle technical challenges. They ultimately decided to employ an air-cushion direct-discharge dual-channel slurry balance shield machine.

This customized shield machine is equipped with a slurry circulation system and a fully pneumatic pressure regulation device for the air cushion chamber, which balances and buffers the soil and water pressure at the foremost excavation face. This works in conjunction with synchronous two-component grouting technology to control surface settlement. Additionally, the shield machine is configured with a composite six-spoke heavy-duty cutterhead. It employs methods like "separated rolling and scraping + targeted tool design" to prevent the formation of a mud cake on the cutterhead, ensuring comprehensive construction safety.

5.3.3 Settlement Control and Grouting Technology

Surface settlement control is a critical issue in tunnel construction. Guangzhou Metro has developed various advanced settlement control technologies, including synchronous two-component grouting technology, the application of quick-setting grout, and precise control of grouting parameters.

In the branch line project of Line 8's northern extension, the builders optimized synchronous grouting by adopting quick-setting grout, controlling the grouting volume and pressure to promptly fill the tail void, thereby preventing ground settlement or shield attitude deviation. Simultaneously, they utilized laser guidance systems and gyroscopes for real-time monitoring of the shield's attitude, ensuring the safe completion of the tunneling task.

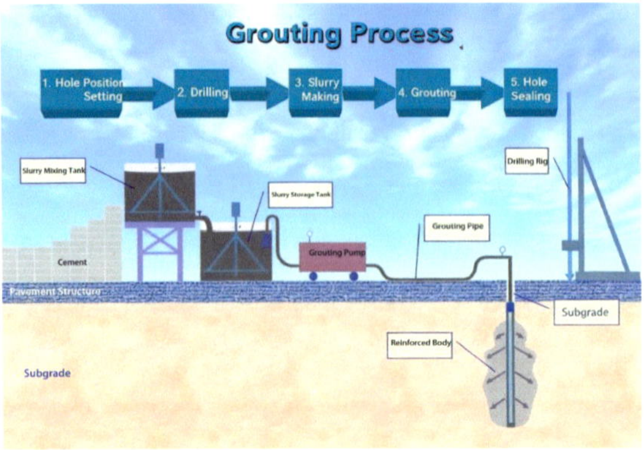

Figure 5-1 Synchronous Two-Component Grouting Technology

5.3.4 Slurry Circulation System and Loop Control

For slurry balance shield machines, the normal operation of the slurry circulation system is crucial. The air-cushion direct-discharge dual-channel slurry balance shield machine applied by Guangzhou Metro is equipped with an advanced slurry circulation system and a fully pneumatic pressure regulation device for the air cushion chamber, which can effectively balance and buffer the soil and water pressure at the foremost excavation face.

This system can automatically adjust the slurry pressure and flow rate based on changes in geological conditions and tunneling parameters, ensuring the stability of the excavation face and preventing surface settlement and machine subsidence.

5.3.5 Emergency Management and Safety Assurance System

Guangzhou Metro has established a comprehensive emergency management and safety assurance system for tunnel construction. In the Line 8 Northern Extension project, to ensure ground safety during shield tunneling, the project department established an above-ground and underground linkage mechanism. Management personnel were arranged for 24-hour ground patrols and reporting, achieving seamless integration of risk early warning and emergency response through bidirectional information sharing.

After the shield launch, the project department set up mobile enclosures on the ground above the shield machine and assigned an emergency vehicle equipped with a generator, grouting equipment, materials, and other supplies to follow the entire process. This ensures that in the event of emergencies like ground settlement, the dangerous area can be immediately isolated, and grouting reinforcement operations can be carried out simultaneously to minimize safety risks.

5.4 International Application and Export of Tunnel Construction Technology

With the deepening of the Belt and Road Initiative, Guangzhou Metro's tunnel construction technology has begun to go global, providing Chinese solutions and wisdom for international rail transit construction.

5.4.1 Application in Belt and Road Projects

Guangzhou Metro's tunnel construction technology has been applied in several Belt and Road partner countries. For example, Guangzhou Metro Engineering Consulting Co., Ltd. secured its first overseas Belt and Road project – the Supervision Project for the Electric Multiple Units of the Lagos Light Rail Blue Line in Nigeria – and completed the train delivery within 7 months. Furthermore, it won the first Guangzhou Metro overseas project, the Macau Light Rail East Line Design-and-Build Project - Project Management and Technical Assistance Services, with a contract value of 338 million Macau Patacas.

5.4.2 International Technical Cooperation and Exchange

Guangzhou Metro actively engages in international technical cooperation and exchange, promoting the internationalization of Chinese tunnel construction technology. In October 2024, Guangzhou Metro and Singapore's SMRT Trains Ltd signed a Memorandum of Understanding. Both parties explicitly agreed to strengthen cooperation and exchange focusing on rail transit operation and maintenance, particularly in metro station energy management, metro tunnel inspection and monitoring, track facility renewal, and full lifecycle system management of rail transit based on safety, reliability, and intelligence.

This international cooperation not only helps promote Guangzhou Metro's advanced technologies but also allows for the absorption of international advanced experience, further enhancing technical levels and service quality.

5.4.3 Academic Exchange and Standard Setting

Guangzhou Metro actively participates in international academic exchanges and standard-setting work. Through international organizations like the International Tunnelling and Underground Space Association (ITA), Guangzhou Metro's technological innovations have gained international recognition. For instance, the "TBM + EPB + Slurry" triple-mode shield machine used in Guangzhou Metro Line 7 Phase 2 won the ITA Annual Product/Equipment Innovation Award, becoming one of three award-winning projects from China.

This international recognition not only enhances Guangzhou Metro's international influence but also provides strong support for bringing Chinese tunnel construction technology to the world.

5.5 Vocational Education and Talent Cultivation in Tunnel Construction

Technological innovation and talent cultivation are mutually reinforcing. While innovating in tunnel construction technology, Guangzhou Metro also places great importance on vocational education and talent cultivation.

5.5.1 Construction of a Vocational Education System

Guangzhou Metro has built a complete vocational education system for tunnel construction, including multiple levels such as professional technical training, practical skills training, and international certification courses. Through innovation platforms like the Guangdong Provincial Engineering Technology Research Center for Shield Tunneling in Composite Strata and the Guangzhou Rail Transit Shield Technology Research Institute, it has a team of over 300 domestic and international shield experts, providing strong teaching resources for talent cultivation.

5.5.2 International Talent Cultivation

Guangzhou Metro focuses on cultivating international talents. Through international project practice, overseas training, and international cooperation exchanges, it has cultivated a group of interdisciplinary professionals proficient in both technology and international rules. For example, during the successful expansion into overseas and international markets by Guangzhou Metro Engineering Consulting

Co., Ltd., a group of professionals with international perspectives was cultivated.

Chapter 6:Station Structure and Architectural Technology

As critical nodes in the urban rail transit network, metro stations serve not only as functional transportation spaces but also as windows showcasing the city's image and integral components of civic life. Throughout its nearly three decades of development, Guangzhou Metro has achieved remarkable accomplishments in the field of station structure and architectural technology, forming a comprehensive technical system and a distinct design philosophy. This chapter will systematically elaborate on the practical experience of Guangzhou Metro stations in structural engineering, architectural design, functional layout, and technological innovation, and explore their application value in international vocational education.

6.1 Innovation and Technological Development in Station Structure Engineering

The structural engineering of Guangzhou Metro stations has undergone a development process from simplicity to complexity and from tradition to innovation. In facing the challenges posed by Guangzhou's complex geological conditions and dense urban environment, a series of technological breakthroughs have been achieved.

6.1.1 Prefabricated Station Technology System

Prefabricated construction technology is a major innovation in the construction of Guangzhou Metro stations. The Shangyong Park Station on Guangzhou Metro Line

11 is the first underground three-story prefabricated station in China and the first prefabricated station in South China. It spans 221.7 meters in length and 22.3 meters in width at the standard section. The construction of this station adopted several innovative technologies:

In terms of construction techniques, the project team employed twin-wheel milling trenching technology. By replacing conventional gears with combined cutter-cone gears, the issue of claystone sticking to the wheels was effectively resolved. Additionally, the equipment deviation correction system and ultrasonic testing were utilized to control the verticality and flatness of the trenches, improving construction precision and efficiency. Structurally, the team developed a technology that integrates the diaphragm wall of the prefabricated metro station as the main structural side wall, combining temporary support structures with permanent ones. This enhanced the overall structural integrity and construction quality.

To meet the precision requirements for installing key components, the team adopted the "full-rotation drilling rig + flange plate positioning" process, achieving a vertical accuracy of 0.2% for steel pipe columns, which is an industry-leading level. Furthermore, an intelligent hoisting device for prefabricated components was developed, featuring precise micro-adjustment, electrical anti-sway, and fixed-point operation functions. This enabled efficient and accurate installation of large-area prefabricated panels.

Figure 6-1 Shangyong Park Station Three-story Prefabricated Station

6.1.2 Application of Ultra-High Performance Concrete

In the construction of Pengbian Station on the second phase of Guangzhou Metro Line 14, Ultra-High Performance Concrete technology was innovatively adopted, marking the first application of UHPC in the internal structure of a metro station in China.

Compared with traditional cast-in-place platform slabs, UHPC prefabricated platform slabs offer multiple advantages: They integrate green technologies, achieving standardized integration, industrialized production, and assembled construction. This method saves one-third of the time compared to traditional methods, reduces required concrete materials by approximately 59%, effectively decreases construction pollution, and improves production efficiency as well as safety and quality levels. The application of this material represents the trend towards greener and more industrialized development in metro station construction.

6.1.3 Integrated Permanent and Temporary Structure Technology

Henan Station on the second phase of Line 14 is the first station in Guangzhou Metro constructed using the open-cut method with an integrated permanent and temporary structure. The station is 155 meters long, designed as a three-story underground structure, with a standard section excavation depth reaching 24.13 meters. The construction adopted the "Open-cut Top-down Construction Method with Integrated Permanent and Temporary Structures." Through systematic design innovation, the temporary support system was organically integrated with the permanent structure.

This technology innovatively combines the concrete bracing system of the foundation pit with permanent components of the main station structure. Station mezzanine slabs and roof slabs replaced the traditional temporary support structures, successfully avoiding construction waste generated from the demolition of temporary supports and effectively reducing building material consumption. This concept of integrating permanent and temporary structures reflects Guangzhou Metro's active exploration in sustainable development.

6.1.4 Station Construction Technology under Complex Geological Conditions

Guangzhou is often called the "Geological Museum" due to its complex and variable geological conditions, which pose severe challenges to metro station construction. Ersha Island Station on Guangzhou Metro Line 12 is located on a river island with complex geology. Construction faced challenges such as difficulties in quality control of trenching in sand layers, a sensitive surrounding environment, and restrictions on muck transportation.

Constructors optimized the construction scheme by eliminating slot wall reinforcement on both sides of the diaphragm walls and summarized prior construction experience to enhance efficiency. Regarding pipeline impacts, they planned ahead, optimized resource allocation and process parameters. After three and a half months of intensive work, they successfully completed the construction of 211 diaphragm wall panels, finishing the enclosure structure 35 days ahead of the original schedule.

During the construction of Chigang Station on Line 12, facing the challenge of being the deepest station on the entire line (with a depth of 41 meters, comparable to a 13-story underground building), the constructors replaced the traditional wire mesh and shotcrete slope protection with the fabric formwork concrete process. This method both prevents river water scouring and protects the river channel environment. They innovatively employed the "cover-and-excavate method" close-fitting technology, constructing six new pillars to rigidly support the existing structure. This allowed for "zero-distance" construction both over and under the existing operational metro line, solving the challenging problem of high-risk underground excavation.

Table 6-1 Innovative Application Cases of Guangzhou Metro Station Structure Technology

Technology Type	Application Station	Technical Features	Innovation Value
Prefabricated Station Technology	Shangyong Park Station, Line 11	Three-story underground prefabricated structure, Intelligent hoisting of prefabricated components	First of its kind in China; Increases construction efficiency by over 30%
Ultra-High	Pengbian Station,	Reduces material usage by	First application in

Performance Concrete (UHPC)	Line 14	59%, Increases construction efficiency by 1/3	China for metro station internal structures; Green and environmentally friendly
Integrated Permanent & Temporary Structure	Henan Station, Line 14	Integration of temporary and permanent structures, Reduces construction waste	First open-cut station using integrated permanent and temporary structure technology
Construction in Complex Geology	Ersha Island Station, Line 12	Trenching control in sand layers, Pipeline protection technology	Addresses construction challenges on a river island

6.2 Station Architectural Design and Humanistic Care

Guangzhou Metro station architectural design not only focuses on functionality and safety but also emphasizes humanistic care and the expression of regional culture, forming a unique design philosophy and practical experience.

6.2.1 International Design Style

Guangzhou Metro Line 18 and Line 22 represent the design concept of a new generation of metro stations. Line 18 stations adopt an overall streamlined aesthetic design, creating agile and elegant layers as well as expansive and smooth spatial perceptions. Line 22 stations utilize diamond elements inspired by diamond cutting, pursuing extreme detail in architectural craftsmanship and modern, fashionable material textures.

The station spaces of these two lines employ a 全新的 design concept, comprehensively creating iconic station spaces that showcase the new image of Guangzhou Metro and meet the characteristics of the new era of rail transit. The entire lines use a material palette of gray and white, highlighting metallic textures and the contrast between solid and void, while punctuating with vibrant line colors, creating an elegant and composed, simple yet fashionable atmosphere. Line 18 uses "Bay Area Blue" as its line color, and Line 22 uses "Vibrant Orange," which also correspond to the regional positioning and future development of the respective lines.

6.2.2 Integrated Design Concept

Guangzhou Metro station design emphasizes an integrated design concept. Starting from the station's architectural space, it fully applies the principles of spatial integration to maximize the use of station space, integrate equipment endpoints, focus on enhancing passengers' spatial perception and comfort, create open and grand visual effects, and reflect the humanized design of the spatial environment.

Simultaneously, designers treat all elements within the station space as a whole for design, considering both their individual characteristics and their interrelationships, ensuring mutual coordination and inclusion. This integrated design concept makes the station space more unified and harmonious, enhancing the overall spatial quality.

6.2.3 Expression of Regional Culture

Guangzhou Metro highly values the integration of station design with regional culture. In the design of Line 6, culturally landmark stations such as Cultural Park, Huangsha, Haizhu Square, Beijing Road, Dongshankou, and Huanghuagang were selected. Large "Cultural Walls" in the concourses display Guangzhou's historical culture, including Xiguan Grand Houses, Dongshan Mansions, the Millennium Commercial Capital, and revolutionary history, designing Line 6 as Guangzhou's first metro line dedicated to promoting historical and cultural landmarks.

This philosophy of integrating local culture into station design is also reflected in Guangzhou Metro's external technical services. For example, the Guangxi First Metro Line – Nanning Line 1, for which Guangzhou Metro Design Institute was responsible for the overall design, features station finishes and entrance/exit designs themed "Qingxiu Yongzhou," extensively adopting elements like Zhuang brocade, fully embodying the ethnic characteristics and style of Guangxi.

6.2.4 Elevated Station Design Innovation

The elevated station design for Guangzhou Metro Line 21 also reflects innovative concepts. The four elevated stations (Changping Station, Jinkeng Station, Zhucun Station, Shantian Station) adopt unified design elements to form a consistent station image, enhancing the recognizability of the line's stations.

The station design starts with the concept of "Simplicity," simplifying complexity, and organically integrates green and environmentally friendly energy-saving measures, striving to create simple, technological, and highly recognizable stations. The platform level design, based on the Lingnan region's climate and the station's surrounding environment, integrates functions such as ventilation, sun protection, and lighting with the architecture. Elevated treatment is used to form open and transparent platform spaces, utilizing outdoor airflow for platform ventilation and introducing the outdoor environment into the building.

Jinkeng Station also implements a series of energy-saving measures such as rainwater collection, permeable paving, sunken green spaces, and photovoltaic power generation, creating a low-energy consumption transportation building focused on greening and environmental protection, and it has obtained the Three-Star Green Building Design Label.

6.3 Station Functional Layout and Passenger Experience Optimization

Guangzhou Metro consistently adopts a passenger demand-oriented approach in station functional layout, continuously optimizing circulation routes and service facilities to enhance the passenger travel experience.

6.3.1 Intelligent Service Systems

Guangzhou Metro has launched a new wayfinding signage system on Lines 18 and 22. Suspended equipment and facilities are largely absent from these stations. Concourses feature integrated informational guidance incorporating clocks and public address systems for greater intuitiveness. White guide signs, complemented by lighting, present a minimalist and modern aesthetic. Platform display screens are embedded directly into the platform screen doors, reflecting meticulous attention to detail throughout.

Furthermore, Guangzhou Metro has introduced professional lighting environment design. Utilizing indirect lighting and a combination of multiple lighting forms, station illumination becomes softer and more comfortable, enhancing the sense of depth within the station architecture. An intelligent lighting control system is employed to manage and adjust color temperature and hue, enabling the creation and customization of artistically expressive lighting scenes as needed. This imbues the stations with a heightened sense of futurism and technological sophistication.

6.3.2 Accessibility and Humanized Design

Guangzhou Metro places strong emphasis on accessible design, providing comprehensive barrier-free facilities within stations. These include elevators, tactile paving, wheelchair ramps, and more, ensuring all passengers can conveniently utilize metro services. Concurrently, Guangzhou Metro continually refines the design of station restrooms and nursing rooms to improve the standard of humanized services.

In the design of the elevated stations on Line 21, facilities such as air-conditioned waiting areas and seating are provided on the platforms, offering passengers a superior waiting environment. This humanized design embodies Guangzhou Metro's passenger-centric service philosophy.

6.3.3 Optimization of Transfer Station Design

As critical nodes within the metro network, the design of transfer stations directly impacts line operational efficiency. Guangzhou Metro Line 11, functioning as a circular line, comprises 31 stations, 26 of which are transfer stations, rightfully earning the title "King of Transfers." In designing these stations, Guangzhou Metro prioritizes transfer efficiency and passenger flow organization, achieving seamless transfers through rational spatial planning and circulation design.

For instance, Wufeng Station facilitates an L-shaped transfer with Line 10. Jiangtai Road Station connects with the existing Line 2, where transfer corridors are situated at the eastern end and southeastern side of the concourse respectively, enabling efficient inter-line passenger movement. Yangang Station employs a new structural form as a large-span column-free station. This design allows for more flexible arrangement of escalators and elevators, creates an open and spacious concourse visual environment, facilitates utility routing, and ultimately enhances overall station functionality.

6.4 Green Building Technology and Sustainable Development

Guangzhou Metro actively implements green building technologies in station construction to promote sustainable development of metro infrastructure.

6.4.1 Application of Energy-Saving Technologies

Guangzhou Metro Design Institute maintains close oversight of domestic and international advancements in energy-saving technologies through multiple research initiatives. The experimental phase for regenerative braking energy inversion feedback systems has been completed, while development continues on supercapacitor energy storage applications.

In power supply innovations, research on new low-loss transformer technology has been piloted on Line 7. For station cooling systems, development is underway for high-efficiency chilled water plants, direct expansion systems, and buried cooling towers. A notable example is the cooling station at Haizhu Square Station on Line 6, which innovatively utilizes direct Pearl River water cooling. This approach eliminates cooling towers while providing cooling sources for adjacent stations, demonstrating significant energy conservation, environmental benefits, and social value.

6.4.2 Eco-Friendly Materials and Construction Techniques

Guangzhou Metro proactively adopts environmentally friendly materials and techniques to minimize construction impacts. The Phase II project of Line 14 introduced a novel "pre-derusting, grinding, and welding" method for rails, embedding them into monolithic track beds to improve wheel-rail contact and reduce vibration and noise at both source and transmission paths.

Furthermore, the metro system extensively implements LED lighting, intelligent lighting controls, light guide illumination, and solar photovoltaic technologies to substantially reduce energy consumption. In designing elevated stations for Line 21, photovoltaic roofs were incorporated as active energy-saving measures. These not only harness solar energy to reduce station power consumption but also integrate seamlessly with canopy designs, creating cohesive architectural fifth facades.

6.5 International Application and Vocational Education Value of Station Construction Technology

The innovative achievements of Guangzhou Metro in station structure and architectural technology not only serve local metro construction but also extend to the international market through technology transfer and vocational education.

6.5.1 Technology Transfer and International Application

The business scope of Guangzhou Metro Design Institute has expanded to 28 cities across China, with participation in metro design projects in multiple global cities. During technology transfer, Guangzhou Metro emphasizes the integration of local experience with regional cultural characteristics. For instance, the design of Nanning Line 1 deeply incorporates local features, extensively adopting elements such as Zhuang brocade to reflect the ethnic characteristics and style of Guangxi.

Guangzhou Metro actively participates in rail transit construction in Belt and Road Initiative partner countries, applying its station design and construction technologies to international projects. For example, Guangzhou Metro Engineering Consulting Co., Ltd. secured its first overseas Belt and Road project—the Supervision Contract for Electric Multiple Units of the Lagos Light Rail Blue Line in Nigeria—and completed train delivery within seven months. Additionally, it won the first Guangzhou Metro overseas project—the Project Management and Technical Assistance Service Contract for the Design and Build of Macao Light Rail East Line.

6.5.2 Vocational Education and Talent Development

Guangzhou Metro's experience in station technology and architectural innovation provides rich materials for international vocational education. By systematically summarizing and refining these technological innovations, Guangzhou Metro has established a comprehensive vocational education system:

In prefabricated station technology, based on the practical experience from Shangyong Park Station, the project team contributed to the development of four industry specifications and acceptance standards. These documents transform innovative achievements and successful practices into shareable technical standards for the industry, providing standardized content for vocational education.

Regarding green building technology, Guangzhou Metro's application cases of energy-saving and environmental protection technologies—such as UHPC materials, integrated permanent and temporary structures, and photovoltaic systems—serve as advanced teaching materials for green building education in rail transit.

In station design, Guangzhou Metro's philosophies of cultural expression, humanized design, and integrated design demonstrate how to balance functional requirements with cultural representation through practical cases. These approaches offer innovative design perspectives for international vocational education.

Table 6-2 Key Areas for International Vocational Education in Guangzhou Metro Station Technology

Technology Area	Core Content	Typical Case	Vocational Education Value
Prefabricated Station Technology	Prefabricated components, Intelligent hoisting, Precision control	Shangyong Park Station, Line 11	Standardized specification export, Leading industry transformation
Green Building Technology	Energy-saving materials, Eco-friendly techniques, Resource recycling	Pengbian Station, Line 14 (UHPC application)	Sustainable development concept implementation
Cultural Expression Design	Regional cultural elements, Spatial artistic expression	Line 6 Cultural Wall, Nanning Line 1	Cultural heritage and innovative design methodology
Intelligent Service Systems	Wayfinding signage, Lighting environment, Information services	Intelligent systems of Lines 18 & 22	Smart station construction and management

Guangzhou Metro has accumulated substantial practical experience and achieved significant technological innovations in the fields of station structure and architectural technology. Ranging from prefabricated station construction and green building technologies to cultural expression and intelligent services, a comprehensive technological system and design philosophy have been established. These achievements not only support the high-quality development of Guangzhou Metro but also provide valuable resources for the international dissemination of China's urban rail transit technology and for vocational education.

Looking ahead, with the continued advancement of the Belt and Road Initiative and the rapid development of global urban rail transit, Guangzhou Metro's expertise in station technology and architectural innovation will play an increasingly important role in broader international cooperation and vocational education. By integrating technological innovation with vocational training, Guangzhou Metro is contributing Chinese expertise and solutions to the development of urban rail transit systems worldwide.

Chapter 7:Track and Power Supply System Construction

The track and power supply systems serve as the "blood vessels" and "nervous system" of metro operations, directly determining the safety, efficiency, and sustainability of rail transit. Throughout its nearly three decades of development, Guangzhou Metro has established a technologically influential system in areas such as track structure, power supply configuration, and intelligent operation and maintenance, achieved through continuous technological innovation and engineering practice. This chapter systematically reviews Guangzhou Metro's key technological breakthroughs, typical application cases in track and power supply system construction, and their pathways for integration into international vocational education, thereby providing a replicable "Guangzhou Solution" for the global rail transit industry.

7.1 Track System Technological Innovation: From Basic Structure to Intelligent Maintenance

7.1.1 Embedded Continuous Supported Track System: Breaking Through Traditional Vibration Damping Bottlenecks

Guangzhou Metro applied the Embedded Continuous Supported Track System on the Knowledge City Branch Line of Line 14, filling a technological gap in China's metro sector. This system replaces traditional fastenings with elastic pads and polymer damping materials, achieving continuous support and locking of the rails, thereby suppressing vibration and noise at the source. Compared to traditional tracks, its vibration damping performance is improved by over 30%, noise is reduced by 5-8 dB, and routine maintenance workload is decreased by approximately 40%. Recognized with a first-class scientific and technological progress award from the China Association of Metros, this technology provides an effective solution for addressing environmental challenges on high-density operational lines.

7.1.2 Trackless Construction Technology: Enhancing Adaptability in Complex Environments

During the construction of Line 13, Guangzhou Metro innovatively employed a trackless construction methodology. By integrating data such as rail level elevation and alignment centerline using total stations and BIM technology, the power supply system installation was completed prior to track laying. This approach shortened the construction period by 20% and mitigated safety risks associated with concurrent operations, demonstrating particular suitability for projects involving complex geological conditions or constrained workspaces (e.g., tunnel interiors). This technology has subsequently been implemented in Guangzhou Metro's newer line projects and serves as a valuable reference for track construction in geologically diverse regions along the Belt and Road Initiative.

7.2 Power Supply System Technology System: From Localization Breakthroughs to Intelligent Upgrades

7.2.1 Intelligent Leap in DC Traction Power Supply Systems

The early Guangzhou Metro Line 1 utilized a 1500V DC bilateral power supply system, relying on imported equipment for overhead contact line protection. To break this technological monopoly, Guangzhou Metro collaborated with the 712 Research Institute of China State Shipbuilding Corporation to develop a new generation intelligent DC traction power supply system, which commenced demonstration operation in 2020 at Baijiang Station on Line 13. The core equipment of this system is a fully domestically produced intelligent circuit breaker capable of real-time monitoring of 18 parameters including temperature and mechanical characteristics, achieving a fault prediction accuracy rate exceeding 95% and a short-circuit breaking time under 15 milliseconds (20% shorter than imported equipment). Millisecond-level data via a cloud platform reduces operational maintenance costs by 30% and shortens fault response time from 45 minutes to 3 minutes.

7.2.2 AC 25kV Co-phase Power Supply Technology: Facilitating High-Speed Regional Express Lines

For 160 km/h regional express lines (such as Lines 18 and 22), Guangzhou Metro led the formulation of the national standard "AC 25kV Co-phase Power Supply Equipment for Rail Transit Traction Power Supply Systems" (GB/T 45658-2025), marking the first application of co-phase power supply technology in urban rail transit. This technology compensates for single-phase load imbalance, eliminates electrical phase separation sections, enabling trains to pass through power supply partitions without interruption, increasing operational efficiency by 15% and improving braking energy recovery rates. This technology provides crucial support for cross-city express lines in the Guangdong-Hong Kong-Macao Greater Bay Area and has been exported to projects such as the Fuzhou-Changle Binhai Express Line.

7.2.3 International Leadership in Power Supply Safety Standards

Guangzhou Metro presided over the formulation of the "Safety Control System for Rail Transit Power Supply Operations" (GB/T 45659-2025), which standardizes safety control parameters and inspection rules for power supply operations, covering multiple scenarios including regional railways and intercity railways. The standard integrates requirements from international standards such as EN 50121 and NF F 01-510, emphasizing the reliability of DC/DC module power supplies under wide voltage input conditions (e.g., complying with the RIA12 standard's requirement for 3.5 times surge withstand capability), thereby establishing technical thresholds for equipment stability in high-temperature and high-humidity environments.

7.3 Localization and International Application of Key Technical Equipment

7.3.1 Core Equipment Breaking Foreign Monopolies

• **Current Collector Localization: Zhongtong** Electric's current collector series has replaced products from France's Mersen and Germany's STEMMANN. Deployed on Guangzhou Metro Lines 14 and 21 (speeds exceeding 120 km/h), they have also been exported to projects in Argentina and Pakistan, receiving China's Excellent Patent Award.

• **Large-Scale Application of Intelligent Circuit Breakers: By** 2021, over 200 intelligent circuit breakers had been installed on lines including Guangzhou Metro Line 13, Qingdao Metro Line 4, and Nanning Metro Line 5, covering voltage levels from 750V to 1500V.

7.3.2 Integrated "Going Global" of the Entire Industry Chain

• Guangzhou Metro promotes the international expansion of upstream and downstream enterprises through an industrial alliance model. Examples include:

• The Pakistan Lahore Orange Line Project utilizes Chinese standard DC 750V third-rail power supply, with Guangzhou Dinghan supplying the air conditioning systems and Union Communication providing the passenger information systems.

• Baiyun Electric has participated in power supply projects for over 100 lines in 33 cities globally. Panyu ICT's intelligent systems cover 32 cities, including the Changsha Metro Line 6.

7.4 Intelligent Maintenance and Safety Assurance System

7.4.1 Intelligent Maintenance Platform for Power Supply Systems

The intelligent power supply maintenance platform developed by Guangzhou Metro integrates BIM and GIS data, enabling predictive maintenance of equipment status. For instance, by analyzing trends in the mechanical characteristic parameters of circuit breakers, potential faults can be predicted 30 days in advance, preventing unplanned outages.

7.4.2 Innovations in Track Monitoring and Maintenance

• **Integrated Boulder Blasting Technology: Addressing** boulder obstructions within shield tunnels, this technology involves geological exploration and precision blasting techniques, ensuring track alignment accuracy. It has received a first-class

Scientific and Technological Progress Award.

- **Rail Pre-treatment Process:** The second phase of Line 14 adopted an integrated solution of pre-derusting and welding rails, controlling vibration and noise at the source and extending track service life.

7.5 International Vocational Education Output Pathways for Track and Power Supply Systems

7.5.1 Development of Standardized Curriculum Systems

Based on Guangzhou Metro's technological achievements, vocational education modules centered on Chinese standards have been constructed:

Basic Courses: Fundamentals of DC Power Supply Systems, Trackless Construction Technology.

Advanced Courses: Co-phase Power Supply Equipment Design, Intelligent Circuit Breaker Operation and Maintenance (incorporating the Baijiang Station case study).

Safety Certification: International certification training for the Safety Control System for Power Supply Operations (GB/T 45659-2025).

7.5.2 Establishment of Overseas Project Training Bases

- **Pakistan Lahore Orange Line Training Center:** Provides DC third-rail power supply maintenance training for local staff, promoting an integrated "theory + practice" model.
- **Guangzhou Metro National Engineering Laboratory:** Collaborates with the Hong Kong University of Science and Technology to jointly cultivate PhD students, conducting research on power supply system big data analysis and AI-driven maintenance.

7.5.3 Multilingual Conversion of Technical Standards

National standards like GB/T 45658-2025 have been translated into English and Spanish versions to meet the needs of target markets such as Argentina and Malaysia. Simultaneously, a multilingual VR teaching system has been developed to simulate power supply equipment debugging processes, reducing cross-border training costs.

7.6 In-Depth Case Study: Integrated Innovation in the Guangzhou Metro Line 18 Power Supply System

As a 160 km/h regional express line, Line 18 fully utilizes AC 25kV co-phase power supply technology, achieving three major breakthroughs:

Continuous Power Supply: Elimination of neutral sections enables trains to receive power uninterrupted throughout the journey, enhancing operational efficiency.

Energy Recovery: Braking energy is fed back into the grid, reducing energy consumption by 10%.

Intelligent Protection: The adoption of an adaptive current gradient algorithm makes short-circuit fault isolation speed internationally leading.

This line has become a benchmark for Guangzhou Metro's technology export, attracting study tours from countries and regions including Singapore and Vietnam.

Conclusion

Through technological iteration in track and power supply systems, Guangzhou Metro has achieved a leap from "import dependence" to "localization leadership." Future efforts should focus on further promoting mutual recognition between Chinese and international standards (e.g., EN, IEC), deepening research on the adaptability of intelligent maintenance technologies to the geological conditions along the Belt and Road routes, and cultivating localized technical talent through vocational education, thereby facilitating the global implementation of Chinese rail transit technology.

Chapter 8:Green Construction and Sponge Station Technology

Against the backdrop of accelerating global urbanization, the greening and sustainability of rail transit systems have become crucial indicators of urban modernization. As a leading enterprise in China's urban rail transit sector, Guangzhou Metro has consistently and deeply integrated the concepts of green construction and sponge city into the entire life cycle of metro planning, design, construction, and operation. Green construction aims to minimize the negative environmental impact of metro development through resource-efficient, environmentally friendly, and smart, efficient engineering practices. Sponge stations enhance resilience against extreme weather events via systematic stormwater management and ecological design. Together, they form a vital technological pathway for Guangzhou Metro to achieve its "dual carbon" goals and promote sustainable urban development. This chapter will systematically elaborate on the key technological systems, engineering practices, and application outcomes of Guangzhou Metro in green construction and sponge station development, providing referenceable and replicable technical models for international vocational education.

8.1 Green Construction Philosophy and Overall Framework

Guangzhou Metro implements the core principles of "Energy Saving, Material Saving, Water Saving, Land Saving, and Environmental Protection" (Four Savings and One Environmental Protection) in its green construction practices, establishing a comprehensive green construction framework supported by "Objectives - Technology - Management - Evaluation." This framework emphasizes the efficient utilization of resources, effective control of carbon emissions, and minimization of environmental disturbance throughout the entire project lifecycle. It translates green concepts into concrete technical measures and management processes through standardization and informatization.

At the objective level, Guangzhou Metro has set a series of quantifiable targets, such as a construction waste reuse rate of ≥30%, noise control compliance rate at construction sites of ≥95%, and non-traditional water source utilization rate of ≥20%. At the technological system level, a diversified integrated system encompassing green planning and design, green construction, environmental protection coordination, and smart construction has been formed. At the management system level, relying on an engineering organization model led by the Guangzhou Metro Group with multi-unit collaboration, green construction requirements are incorporated into the entire process of tendering, design, and acceptance. The evaluation system, combining the national "Assessment Standard for Green Building" with the characteristics of metro engineering, formulates green performance assessment criteria for the entire project lifecycle.

Table 8-1 Key Performance Indicators for Green Construction in Guangzhou Metro

Category	**Specific Indicator**	**Target Value**	**Evaluation Method**
Energy Consumption	CO2 emissions per unit passenger turnover	Annual reduction of 3%	Carbon footprint accounting
Material Utilization	Construction waste reuse rate	≥50%	On-site measurement and statistics

Water Resource Utilization	Non-traditional water source utilization rate	≥20%	Water usage ledger analysis
Environmental Protection	Noise control compliance rate at construction sites	≥95%	On-site monitoring

The establishment and implementation of this framework has positioned Guangzhou Metro at the forefront of green construction nationwide, while also laying a solid foundation for the refinement and deepening of subsequent technical systems.

8.2 Key Technical System for Green Construction

8.2.1 Energy Saving and Renewable Energy Utilization

As a major urban energy consumer, Guangzhou Metro significantly reduces energy consumption through three primary pathways: equipment energy efficiency, system optimization, and renewable energy integration.

(1) High-Efficiency Equipment and System Optimization. Given Guangzhou's hot and humid climate, station ventilation and air conditioning systems are key focuses for energy conservation. Projects like Lines 14 and 10 have widely adopted variable frequency drive (VFD) technology, dynamically adjusting fan and pump speeds based on passenger flow and indoor/outdoor temperature and humidity, achieving system energy savings of 20%–30%. Concurrently, elevators and escalators are equipped with energy regeneration units that feed regenerative power generated during braking back into the station grid, achieving energy savings of 15%–25%. For lighting, the entire network utilizes LED intelligent lighting systems, enabling zoned and timed lighting through light sensing and time control, resulting in over 60% energy savings compared to traditional lighting.

(2) Renewable Energy Integration. Guangzhou Metro deploys solar photovoltaic (PV) systems on a large scale using roof resources at depots, stabling yards, and some elevated stations. For example, the Yuzhu Depot PV project generates over one million kilowatt-hours annually, effectively reducing the operational carbon footprint. Furthermore, ground-source heat pump systems are being piloted in suitable new stations, utilizing the constant temperature of underground soil for station heating and

cooling, thereby further optimizing the energy mix.

8.2.2 Material Saving and Material Resource Recycling

To reduce building material consumption and construction waste generation, Guangzhou Metro has developed a systematic technical framework covering material selection, process innovation, and resource recycling.

(1) Green Building Materials and Prefabricated Assembly: In network construction, Guangzhou Metro prioritizes products certified as green building materials, such as low-VOC (Volatile Organic Compound) paints, ready-mix concrete, and ready-mix mortar, controlling pollution and waste at the source. More significantly, Guangzhou Metro has achieved major breakthroughs in prefabricated assembly technology. For instance, Pengbian Station on Line 14 uses Ultra-High Performance Concrete (UHPC) prefabricated platform slabs, requiring approximately 59% less concrete material than traditional cast-in-place methods, increasing construction efficiency by one-third, and significantly reducing on-site construction pollution.

(2) Construction Waste Reduction and Resource Recovery: Guangzhou Metro implements a full-chain waste management model of "source reduction - process control - end utilization" during construction. This involves precise setting-out, factory prefabrication, and digital fabrication to reduce on-site waste; separate collection of waste concrete, bricks, and steel; and the establishment of construction waste resource recovery centers to convert waste into recycled aggregates for use in roadbed filling and site hardening. In the Line 18 and 22 projects, the construction waste reuse rate has exceeded 50%.

8.2.3 Water Saving and Water Resource Recycling

Guangzhou Metro comprehensively enhances water resource utilization efficiency through the widespread use of water-saving fixtures, development of non-traditional water sources, and intelligent management.

(1) Water-Saving Fixtures and Smart Monitoring: All stations across the network are equipped with the highest water efficiency grade sanitary fixtures, including sensor-activated taps and water-saving toilets. Simultaneously, pilot installations of

smart water meters and leakage monitoring systems in hub stations enable real-time analysis of water usage data, allowing for prompt detection and repair of pipeline leaks.

(2) Non-Traditional Water Source Utilization: During metro foundation pit construction, abundant groundwater inflow is treated through sedimentation and filtration and then used for dust suppression, vehicle washing, and concrete curing, significantly reducing municipal water consumption. Some stations are also designed with rainwater harvesting systems that collect and purify rainwater from station building and yard roofs for use in landscape irrigation and station cleaning, achieving water resource recycling.

8.2.4 Integrated Permanent-Temporary Structures and Green Construction Methods

"Integrated Permanent and Temporary Structures" is another innovative concept in Guangzhou Metro's green construction, focusing on the organic integration of temporary facilities and permanent structures during the construction phase to avoid redundant construction and resource waste.

Henan Station on Line 14 is an exemplar of this philosophy. The station adopted the "Open-cut Top-down Construction Method with Integrated Permanent and Temporary Structures," merging the design of the temporary concrete bracing system for the foundation pit with permanent components of the main station structure. This approach successfully avoided the construction waste generated from demolishing temporary supports in traditional methods, while also reducing the consumption of building materials like steel and concrete, achieving multiple benefits in terms of sustainability, economy, and construction efficiency.

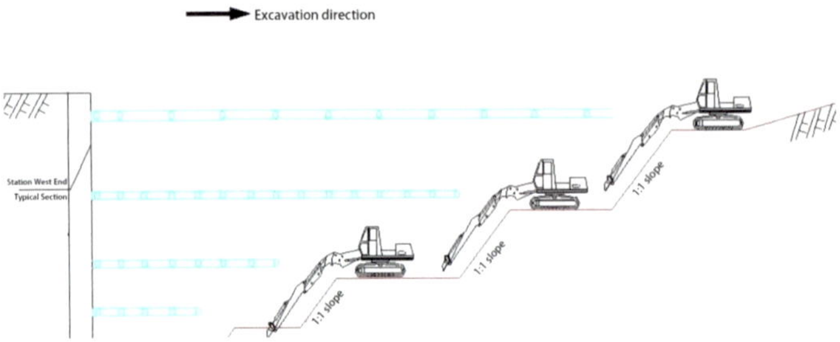

Figure 8-1 Open-cut Top-down Construction Method

Table 8-2 Typical Cases of Green Construction Technology Application in Guangzhou Metro

Technology Category	Typical Project	Technical Features	Implementation Outcomes
HVAC Energy-saving Retrofitting	Environmental Control System Retrofitting for Lines 3 & 5	Adoption of ultra-high efficiency intelligent environmental control system; retrofitting of 42 stations	Expected electricity savings of 450 million kWh during contract period; average energy saving rate of 41.5%
Prefabricated Assembly Technology	Pengbian Station, Line 14	UHPC prefabricated platform slabs	Material usage reduced by 59%; construction efficiency increased by 33%
Photovoltaic Power Generation System	Guangzhou New Town Stabling Yard, Line 3 Eastern Extension	Roof-mounted solar photovoltaic panels	Annual power generation of 108,000 kWh
Mechanized Construction	Line 10, Guangfo-Huadi Intercity Railway	Mechanized method replacing the mining method, reducing open-cut excavation	Resolved pain points in central urban areas such as traffic management and utility relocation

8.3 Key Technical System for Sponge Stations

Sponge Stations represent the concrete application of the Sponge City concept in metro engineering. Through various technical measures encompassing "infiltration, detention, storage, purification, utilization, and drainage," Guangzhou Metro has established systems for the natural retention, infiltration, and purification of stormwater within station areas, effectively mitigating waterlogging risks and enhancing water environment resilience.

8.3.1 Ecological Stormwater Management Techniques for Station Areas

(1) Ecological Drainage Facilities: At multiple stations, such as Shangyong Station on Line 11, ecological facilities like permeable pavements, grassed swales, and rain gardens are widely used in station plazas and surrounding green spaces. These facilities effectively promote rainwater infiltration, delay the peak surface runoff, and filter and purify pollutants such as suspended solids and oils from the initial stormwater runoff.

(2) Stormwater Storage Tanks: To cope with extreme rainstorms, Guangzhou Metro installs underground stormwater storage tanks in low-lying areas of stations or near sunken plazas. When rainfall intensity exceeds the drainage capacity of the pipelines, these tanks can temporarily store excess stormwater, releasing it slowly into the municipal drainage network after the peak flow has passed, effectively preventing waterlogging and backflow at critical locations like station entrances/exits and ventilation shafts.

8.3.2 Flood Protection Design System and Key Technologies

Flood protection is a fundamental safety requirement for Sponge Station operation. Based on the principle of "multiple defenses and graded protection," Guangzhou Metro has established a comprehensive flood protection technical system covering stations, tunnels, depots, and other facilities.

1.Flood Protection Design for Underground Stations

- **Entrances/Exits: As** the first line of defense, entrance/exits platforms are designed to be ≥0.45 meters above the outdoor ground level and are equipped with flood barriers or intelligent flood gates. Key stations are additionally equipped with automatic water level monitoring and warning systems that can trigger flood protection equipment automatically when water accumulation reaches a threshold.

- **Ventilation Shafts, Air Shafts, and Emergency Exits:** Their opening heights are required to be above the surrounding ground level. Ventilation shafts located by roadsides require an opening height ≥2 meters, while those in green spaces require ≥1 meter. Air shafts and emergency exits have a flood protection height ≥1 meter to prevent rainwater backflow.

- **Sunken Plazas:** These areas are vulnerable points. Flood protection measures with a height ≥1 meter are required at the connection points between sunken plazas and the station, and the plaza's ground level must be ≥0.3 meters higher than the surrounding area.

2.Flood Protection Design for Tunnels and Other Facilities

- **Underground Tunnels: When** passing under rivers or flood zones, flood gates or other anti-flooding measures are installed at both ends of the tunnel section. The flood protection design standard for tunnel ventilation shafts is required to be no less than the 100-year waterlogging level.

- **Elevated Stations and Sections:** The ground-level concourses of elevated stations are required to be situated no lower than the 100-year waterlogging level. Station external windows or curtain walls must be ≥1 meter above the outdoor ground level. The flood protection height for at-grade track bed sections is ≥1.2 meters.

8.3.3 Integration of Green Facilities and Ecological Landscape

Guangzhou Metro deeply integrates sponge facilities with station landscape design, achieving a unity of function and aesthetics. For instance, at some stations on Line 14, grassed swales and rain gardens not only manage stormwater but also create attractive station-front landscapes through the use of native plants and micro-topography shaping. Furthermore, in station quality improvement projects, Guangzhou Metro implements the refined streetscape concept of "opening up street corners and revealing green spaces," by removing walls and supplementing greenery, thereby better integrating stations into the urban ecological network.

Depots and Control Centers: The site design elevation is required to be 0.5 meters higher than the average level of the surrounding main roads. The flood protection height for underground garage entrances/exits is ≥1 meter.

Table 8-3 Example Indicators for Sponge Station Construction in Guangzhou Metro

Indicator Category	Specific Indicator	Control Requirement	Applicable Station Types
Runoff Control	Annual Total Runoff Control Rate	70%-85%	Depots & At-grade Stations
Ecological Facilities	Sunken Green Space Ratio	≥15%	Depots & At-grade Stations
Permeable Paving	Outdoor Permeable Surface Ratio	≥40%	Station Plazas & Pedestrian Areas
Storage Capacity	Storage Volume per Unit Impervious Area	≥200 m^3/ha	Key Flood-Prone Areas

8.4 Integration of Green and Smart Technologies with Standardization

Guangzhou Metro places high importance on the integration of green construction, sponge station concepts, and smart technologies, and promotes the consolidation and dissemination of advanced technologies through standardization efforts.

8.4.1 BIM and Digital Empowerment

Building Information Modeling (BIM) technology plays a central role in the green construction practices of Guangzhou Metro. During the construction of Line 10, comprehensive BIM detailed design and modular prefabricated construction were fully implemented. For instance, the prefabrication rate for the Mechanical and Electrical (M&E) engineering at Zhongda South Gate Station on Line 10 reached 95%, with wall and ceiling finishing achieving a near 100% prefabrication rate. For the chilled water plant room installation, BIM technology was used to create a detailed model in advance. Through visualized technical disclosure, the work, which traditionally required over 10 workers for one month, was completed by just 8 workers in 3 days, achieving a unity of efficiency, sustainability, and safety.

8.4.2 Variable-Diameter Shield Tunneling and Integrated Station-Tunnel Green Construction

The "14th Five-Year Plan" National Key R&D Program project led by Guangzhou Metro Group – "Key Technologies and Equipment for Green Construction of Urban Rail Transit Integrating Variable-Diameter Shield Tunneling and Stations" – aims to overcome challenges such as the equipment-stratum-structure interaction effects and the superimposed disturbances from integrated station-tunnel construction. This project is dedicated to establishing a technical system for integrated green construction of variable-diameter shield tunnels and stations, encompassing "engineering design → equipment development → green construction → quality control." It explores advanced construction modes characterized by "small excavation faces, low impact, and minimal interference," representing one of Guangzhou Metro's core future directions for green and intelligent construction.

8.4.3 Standardization and International Dissemination

Standards are crucial for the large-scale application of green technologies. In March 2025, the Rail Transit Professional Committee of the China Engineering Construction Standardization Association (CECSA) was formally established, relying on the expertise of Guangzhou Metro Design Institute. This committee is committed to translating Guangzhou Metro's technical achievements in areas like green construction and sponge stations into industry standards, thereby promoting the international dissemination of China's green rail transit technology system.

Guangzhou Metro also released the first climate action report within China's rail transit industry, which has been iteratively upgraded into an annual sustainability report. It has established green and low-carbon design and evaluation standards covering the entire "station-depot-line" system of rail transit. These standardized documents provide systematic teaching references and technical specifications for international vocational education.

8.5 Recommendations for International Vocational Education Transformation

Transforming Guangzhou Metro's green construction and sponge station technologies into international vocational education resources requires focusing on the structured organization of the knowledge system and the applicability of the teaching content.

8.5.1 Curriculum System Design

It is recommended to build a "four-in-one" curriculum system centered on green construction and sponge station technologies, comprising:

Fundamental Theory Module: Covers concepts of green construction, principles of sponge cities, and lifecycle assessment methods.

Core Technology Module: Provides in-depth explanation of key technologies such as high-efficiency environmental control systems, prefabrication and assembly techniques, ecological stormwater management, and flood protection design.

Case Study and Practice Module: Uses projects like Guangzhou Metro Line 18, Jinkeng Station on Line 21, and the Line 3 Eastern Extension as examples for practical, hands-on teaching.

Standards and Specifications Module: Interprets normative documents like the Green and Low-Carbon Design Standards for Rail Transit and the Sponge City Construction Indicator System.

8.5.2 Training Platform Development

To enhance the effectiveness of vocational education, developing the following training platforms is recommended:

Green Construction Technology Training Platform: Integrates functions like BIM modeling, energy consumption simulation, and material performance testing, allowing trainees to simulate the green construction design and decision-making process.

Sponge Facility Monitoring Platform: Uses sensor networks to monitor the runoff control effectiveness of facilities like rain gardens and permeable pavements, visually demonstrating the principles of sponge technologies.

Intelligent Environmental Control System Operation Platform: Enables trainees to practically operate systems for variable frequency drives, energy consumption monitoring, and intelligent control, mastering skills for optimizing the energy efficiency of environmental control systems.

Conclusion

Through systematic innovation and engineering practice in green construction and sponge station technologies, Guangzhou Metro has successfully established a resource-efficient, environmentally friendly, resilient, and safe model for metro construction and operation. Its technical system not only encompasses specific technical measures for energy/water conservation, material saving, environmental protection, and stormwater management but also extends to systematic work like standard formulation, carbon management, and intelligent maintenance, forming a comprehensive green rail transit solution.

Looking ahead, with the ongoing R&D and application of cutting-edge technologies like "Variable-Diameter Shield Tunneling and Integrated Station-Tunnel Green Construction," and the deep utilization of the Suzhou OS platform in intelligent maintenance, Guangzhou Metro's green technology system will be further refined. This will provide more replicable and scalable "Guangzhou Experience" for the sustainable development of global rail transit. For international vocational education, transforming these mature technical systems and practical cases into educational resources to cultivate a new generation of engineering talents equipped with green concepts and technical capabilities will be a key force in promoting the sustainable development of global rail transit.

Chapter 9:TOD Mode and Urban Integrated Development

9.1 TOD Concept and Overview of Guangzhou's Practice

Transit-Oriented Development (TOD), as an urban development model oriented around public transport, has become a significant paradigm in modern urban planning globally. As a pioneer in implementing this concept, Guangzhou Metro has expanded its connotation from a singular "transportation hub" to an "urban functional hub," establishing a comprehensive development framework characterized by station-city integration, functional synergy, and spatial value enhancement. Within Guangzhou's practice, TOD serves not only as a tool to alleviate traffic congestion but also as a strategic instrument for optimizing urban structure, promoting intensive land use, driving industrial upgrading, and enhancing urban quality.

The evolution of Guangzhou Metro's TOD development has progressed from the 1.0 phase of "Station Construction" to the 2.0 phase of "Station Area Development," then to the 3.0 phase of "Station-City Integration," and has now entered the 4.0 phase of "Urban Network Integration." This progression reflects a shift in focus from "transportation function" to "urban function." Its core objective is to achieve multiple values—including urban spatial restructuring, enhanced economic benefits, improved social services, and optimized environmental quality—through the deep interaction between rail transit and urban development.

Table 9-1 TOD Development Stages and Characteristics of Guangzhou Metro

Development Stage	**Key Characteristics**	**Representative Projects**	**Core Objective**
TOD 1.0 (Station Construction)	Focus on transport function; stations relatively independent from the city	Early metro stations	Meet basic travel demands
TOD 2.0 (Station Area Development)	Moderate development around stations; initial improvement of functional support	Some community-level stations	Enhance land value around stations
TOD 3.0 (Station-City Integration)	Integrated station-city design; highly mixed-use functions	Guangzhou Baiyun Station, Zhujiang New Town Station	Build regional urban centers
TOD 4.0 (Network Integration)	Multi-network integration; city-level functional consolidation	Bay Area rail network	Achieve regional coordinated development

In Guangzhou Metro's TOD practice, the overarching principles of "planning-led, phased development, functional diversity, and spatial integration" have been established. Through the integrated planning and design of rail transit stations and surrounding urban areas, a seamless connection has been achieved between transportation functions and urban activities such as living, working, and leisure. This approach has given rise to multiple vibrant urban nodes, serving as exemplary models for optimizing urban spatial structure in the Guangdong-Hong Kong-Macao Greater Bay Area.

9.2 TOD Planning and Urban Spatial Integration Technology

9.2.1 Multi-level Planning Integration Method

Guangzhou Metro has adopted a three-tier integration method—"Region-City-Station"—in its TOD planning, achieving coordinated development between rail transit and the urban spatial structure.

At the regional level, focusing on the three major metropolitan areas within the Greater Bay Area encompassing 11 cities and a nearly 10,000-kilometer composite mega-network of "metro + intercity railways," Guangzhou Metro conducted 23 thematic studies covering planning, construction, operation, systems, and resources. This regional perspective elevates TOD planning beyond the confines of individual cities, forming a transportation backbone that supports the integrated development of the Bay Area.

At the city level, Guangzhou Metro has established a multi-level rail transit network of "Express Line A + Express Line B + Regular Lines," planning for 53 lines with a total mileage exceeding 2,000 kilometers. This network structure enables differentiated TOD strategies based on line characteristics: express line corridors focus on connecting urban centers and sub-centers, while regular line stations emphasize the enhancement of community services and daily living functions.

At the station level, Guangzhou provides comprehensive guidance for TOD integrated development implementation through the "TOD Development Implementation Rules." Since the 13th Five-Year Plan period, the Guangzhou municipal government has approved 74 station complexes, including 3 national railway stations, 38 intercity stations, and 33 metro stations, with a total site area of

870 hectares and a development scale of 19.57 million square meters. This layered and categorized planning approach enables TOD development to more precisely match the developmental needs of different areas.

9.2.2 Composite Land Use Technology

Composite land use is a core technology of the TOD model. In the land development around stations, Guangzhou Metro achieves intensive and efficient use of construction land through three dimensions: composite ownership, composite function, and composite space.

Regarding composite ownership, the station area of Guangzhou Design Valley serves as an example, where comprehensive district-wide development was achieved through the reorganization and integration of land with complex ownership patterns. This model addresses the common issue of land fragmentation in urban development, creating conditions for holistic area development.

Regarding composite function, the Zhujiang New Town station area organically links and centrally arranges diverse land-use functions, achieving integrated development of station, city, industry, people, and culture. This functional mix not only improves land use efficiency but also creates vibrant, 24/7 urban spaces.

Regarding composite space, the Zhenlong Depot station area achieves land intensification and compatibility through three-dimensional development. By stacking functions vertically, it resolves the issue of multi-functional coexistence that cannot be achieved on a single plane, significantly enhancing land value.

Table 9-2 Comparison of TOD Composite Land Use Models in Guangzhou Metro

Composite Type	Typical Case	Core Strategy	Applicable Conditions
Ownership Composite	Guangzhou Design Valley Station Area	Integration of complex land ownership, coordinated area development	Existing land redevelopment areas
Functional Composite	Zhujiang New Town Station Area	Integration of station, city, industry, people, and	Urban central areas, regional

		cultural functions	centers
Spatial Composite	Zhenlong Depot Station Area	Three-dimensional development, vertical function stacking	Large infrastructure sites like depots and stabling yards

9.2.3 Integrated Station-City Design Technology

Integrated station-city design is the key technology for achieving the fusion of "station" and "city" within the TOD model. In the design of comprehensive hubs like Guangzhou Baiyun Station, Guangzhou Metro has implemented integration technologies across five levels: railway-local coordination, rail connectivity, regional transportation, land development, and operational management.

At the level of railway-local coordination, national rail, intercity rail, metro, and transport connection facilities are planned simultaneously to achieve integrated station-city fusion. This collaborative design breaks down barriers between different transportation modes, providing passengers with a seamless transfer experience.

At the rail connectivity level, Guangzhou Baiyun Station reconfigured the rail transit network, planning the introduction of six metro lines and achieving seamless integration between above-ground and underground spaces. This optimizes the area's value and enhances its overall quality. This multi-line interchange design significantly increases the hub's catchment area and service capability.

At the regional transportation level, a regional road traffic distribution system centered on the hub was constructed, featuring an "elevated distribution ring + grid-and-spoke radial roads." Concurrently, 13.7 kilometers of urban arterial roads connected to and surrounding the station were completed in a single phase. This system effectively mitigates the traffic congestion commonly associated with large-scale transport hubs.

Figure 9-1 Schematic Diagram of Integrated Station-City Design

9.3 Key Technology System for TOD

9.3.1 Transportation Integration Technology

Transportation integration forms the foundational support for the TOD model. In advancing the "Integrated Four-Network System," Guangzhou Metro fosters the development of city cluster metropolitan areas "on the rails" by promoting the integration of the Trunk Railway Network, Intercity Railway Network, Regional (Suburban) Railway Network, and Urban Rail Transit Network.

Guangzhou Metro has established principles of "Shared Corridors, Shared Resources, Multi-Network Coordination; Compatible Standards, Technological Innovation; Coordinated Operating Entities, Established Management Rules and Service Standards." These principles, supported by unified planning, unified standards, and unified management, facilitate the comprehensive integration of the four networks across regions and systems, achieving interconnection, interoperability, inter-operation, and mutual maintenance. This provides an institutional guarantee for the integrated operation of multiple transportation modes.

For cross-system interconnection, cross-modal integration is achieved by adding connecting lines and optimizing the network structure. Examples include the Guangqing Intercity, Foshan-Zhaoqing Intercity, and Guangfo-Jiangzhu Intercity railways, which, through the addition of connecting lines with the Guanghua Intercity and Fosui-Guan Intercity, enable direct 1-hour journeys from Qingyuan, Zhaoqing, and Jiangmen to Guangzhou's core areas. This technical approach integrates intercity railways and the metro into an organic whole rather than isolated systems.

Regarding cross-regional resource sharing, the depth and breadth of resource sharing are continuously expanded through "Coordinated Layout, Hierarchical Sharing, Integrated Stations, Shared Corridors, Centralized Maintenance, and Distributed Operation," and "Leveraging Existing Assets, Utilizing Social Resources, and Efficient Rescue." This resource-sharing model significantly improves facility utilization efficiency and investment returns.

9.3.2 Functional Integration and Spatial Organization Technology

The success of TOD projects largely depends on the rationality of functional configuration and spatial organization. In the Yuesheng Plaza project of the Guangzhou Baiyun Station TOD Comprehensive Transport Hub, Guangzhou Metro implemented the "Vertical City" design concept. The project, with a total floor area of 220,000 square meters, comprises four super-high-rise towers integrating commercial office space, residential units, hotels, retail, and leisure entertainment.

By creating "Four-Ground-Level" commercial spaces and 24-hour night economy scenes, it caters to citizens' dual needs for "fast transit + slow living." This spatial innovation breaks the limitations of traditional single-level commerce, creating multi-dimensional, multi-temporal value spaces.

In the Guangzhou Design Valley station area, the layout of five metro stations precisely covers the Guangzhou Design Valley (industrial core), Huangbian Creative Industry-City Integration Development Area, Lianbian Legal Industry-City Integration Development Area, and Hebian Sci-Tech Industry-City Integration Development Area. This enables talent, resources, and information to rapidly converge and disperse via rail transit. This precise alignment of station placement with industrial functions reflects the deep integration of TOD planning with urban industrial upgrading.

9.3.3 Green and Smart Construction Technology

Green and smart construction are essential features of contemporary TOD projects. During the construction of the Guangzhou Baiyun Station TOD Comprehensive Transport Hub, the project team extensively adopted new materials and techniques, using prefabricated enclosures for functions like sound and thermal insulation. A reclaimed water system was implemented to collect, treat, and reuse construction wastewater, conserving water resources and protecting the environment. Green unitized curtain walls with vertical vision glass and horizontal shadow boxes were used to reduce solar heat gain and improve the indoor environmental quality.

Regarding smart construction, the project implemented an intelligent equipment monitoring system, utilizing IoT and AI-based dust monitoring systems and tower crane visual monitoring systems, achieving all-time, panoramic supervision and automated safety management. These technological applications not only improved construction efficiency but also minimized the impact on the surrounding environment.

Guangzhou Metro is also promoting the application of the "Tencent OS" smart system in rail transit construction. This system, already deployed on Lines 18 and 22, enables features like frictionless fare gates and intelligent passenger flow management. Such smart systems provide technical support for the long-term, efficient operation of TOD projects.

9.4 Typical Practice Cases

9.4.1 Guangzhou Baiyun Station TOD Comprehensive Transport Hub

Guangzhou Baiyun Station is a main passenger station among Guangzhou's "Five Major & Four Auxiliary" stations and the city's first supersized hub planned and constructed according to the modern comprehensive transport hub concept. It handles conventional rail services, accommodates some high-speed rail and Pearl River Delta intercity passenger flow, and integrates metro, intercity rail, long-distance bus, and public transport hubs.

As a key component of the Baiyun Station integrated development, the Yuesheng Plaza project, with a total floor area of 220,000 square meters, includes four super-high-rise towers housing Grade A office space, large-format apartments, and a light-luxury hotel from an international brand. This design achieves vertical integration of transport functions with urban commercial and business functions, serving as a model for "station-city integrated" development.

Leveraging six metro lines and an air-rail intermodal system, the project further strengthens the Greater Bay Area's "1-hour living circle." Passenger volume is projected to exceed 43 million by 2030. Currently, the project has attracted international brands like the Pengruili International Hospital and InterContinental Hotels Group, alongside several leading enterprises, accelerating the formation of a headquarters economy and fashion consumption industry cluster.

9.4.2 Guanhu Depot Over-track TOD Development

The Guanhu Depot over-track development project is one of the first batch of depot over-track development projects with reserved space in the Guangzhou area and the first such project where Guangzhou Metro implemented secondary development. Despite lacking mature experience and relevant policies, Guangzhou Metro completed the entire process, from preliminary feasibility studies during network planning to land preparation and secondary development.

This project holds milestone significance in Guangzhou's TOD development history. It was a key pilot project for metro over-track comprehensive development during the 12th Five-Year Plan period and is a representative example of Guangzhou's TOD 3.0 "Station-City Integration." Its successful experience in implementing master plans, improving municipal infrastructure, driving area development, and generating revenue to subsidize metro construction validates the correctness of the TOD-oriented urban regeneration approach.

The project's positioning is highly aligned with urban planning, making it a strong support for Guangzhou's city strategy. It connects the Zengcheng area with the central city via the metro, addressing the pain point of the separation of "industry and residence." Furthermore, the project comprehensively addresses facilities and supporting services for all-age groups and living scenarios within the residential area, activating the district's development momentum and enhancing the quality of the surrounding urban area.

9.4.3 Helong Subdistrict Multi-station TOD Network

Helong Subdistrict is unique within Guangzhou's subdistricts and towns for having five metro stations in the Baiyun District. These stations include the operational Line 2 Huangbian Station, Line 3 North Baiyun Avenue Station, and the soon-to-open Pengbian, Helong, and Hebian stations on the second phase of Line 14.

These five stations precisely cover the Guangzhou Design Valley (industrial core), Huangbian Creative Industry-City Integration Development Area, Lianbian Legal Industry-City Integration Development Area, and Hebian Sci-Tech Industry-City Integration Development Area. The "One Core, One Ring, Three Areas, Five Stations" industry-city integration framework proposed for Helong Subdistrict in the 14th Five-Year Plan is transitioning from blueprint to reality.

This high-density station layout, precisely matched with industrial functional areas, allows talent, resources, and information to rapidly converge and disperse via rail transit, creating a virtuous cycle where stations and industries mutually reinforce each other. Taking the Guangzhou Design Valley as an example, it has already attracted numerous high-quality headquarters enterprises like CCCC Fourth Harbor Engineering Bureau and Guangdong Provincial Transport Planning and Design Institute, forming a complete industrial ecosystem.

9.5 Vocational Education Transformation Pathways

9.5.1 Curriculum Module Design

Transforming Guangzhou Metro's extensive experience with the TOD model and urban integration into international vocational education resources requires building a systematic curriculum. It is recommended to design the following modules around the three levels of TOD Planning & Design, Development & Implementation, and Operation & Management:

- **TOD Fundamental Theory Module:** Covers the evolution of the TOD concept, core principles, value logic, and economic benefit analysis, illustrated with typical cases from Guangzhou Metro's TOD development stages.

• **Planning Technology Module:** Focuses on technical aspects like multi-level planning integration methods, composite land use technology, and integrated station-city design, using project examples like Guangzhou Baiyun Station and Guanhu Depot.

• **Development Implementation Module:** Covers TOD project development models, investment and financing strategies, benefit distribution mechanisms, and construction management, analyzing Guangzhou Metro's innovative practices in land acquisition and cooperative development.

• **Operation Management Module:** Introduces long-term operational management strategies for TOD projects, including smart operation, service integration, and value maintenance, incorporating case studies of smart systems like Suiteng OS.

9.5.2 Training Platform Development

To enhance the effectiveness of TOD vocational education, developing the following training platforms is recommended:

• **TOD Planning Simulation Platform:** Based on GIS systems and 3D modeling software, creates a virtual urban environment allowing trainees to simulate planning decision-making processes for urban design, traffic organization, and functional layout around TOD stations.

• **Development Evaluation Platform:** Provides financial models and investment analysis tools, enabling trainees to conduct economic feasibility analyses for TOD projects and evaluate the return on investment of different development strategies.

• **Integrated Station-City Design Workshop:** Organizes interdisciplinary teams to develop integrated design solutions for real station sites, fostering comprehensive problem-solving skills.

9.5.3 International Adaptation Recommendations

When disseminating Guangzhou Metro's TOD experience internationally, adaptability to different countries and regions must be considered:

•**Standard and Code Adaptation: Localize** adjustments based on the engineering construction standards and planning regulations of different countries, extracting universally applicable technical principles from Guangzhou's experience.

•**Cultural Habit Integration: Consider** differences in resident travel habits and lifestyles across cultural contexts to make targeted adjustments to TOD functional configurations and spatial designs.

•**Implementation Mechanism Innovation:** Design suitable TOD development implementation mechanisms and policy support systems according to the land systems and investment/financing environments of different regions.

9.6 Conclusion

Through years of TOD practice, Guangzhou Metro has established a full-chain technical system encompassing planning and design, development and implementation, and operation and management, creating numerous urban nodes characterized by station-city integration, well-developed functions, and vibrant activity. Its technical strengths are reflected in multiple aspects, including multi-level planning integration, composite land use, transportation integration, functional fusion, and green and smart construction, providing a rich reference of experience for the integrated development of global urban rail transit and cities.

Looking ahead, with the deepening promotion of the "Integrated Four-Network System" and the development of the Greater Bay Area's rail transit network, the Guangzhou Metro TOD model will further enhance its strategic value, progressing from single-point development to networked development, and from urban nodes to regional hubs, supporting the integration of the Bay Area city cluster. For international vocational education, systematically structuring, theorizing, and curricularizing the successful experience of Guangzhou Metro's TOD to cultivate professionals with capabilities in TOD planning, development, and operation management will be a crucial pathway for promoting sustainable urban development globally.

Part Ⅲ: Operation and Maintenance Technology

Chapter 10:Metro Operation Management System

10.1 Operation Management Overview and Framework System

Metro operation management is a systematic activity centered on ensuring safe and convenient passenger travel. It involves the planning, organization, coordination, and control of all elements within the metro system through scientific organization and advanced technology. As a leading enterprise in China's urban rail transit industry, Guangzhou Metro has established a modern operation management system aimed at "Safety, Efficiency, Intelligence, and Greenness." This system comprehensively covers key operational areas including train operation organization, passenger service, equipment maintenance, safety assurance, and emergency management.

The Guangzhou Metro operation management system adheres to the core principles of "Networked Operations, Standardized Management, Intelligent Enablement, and Humanized Services." Against the backdrop of continuous network expansion, it has achieved a transformation and upgrade from independent single-line operation to coordinated network management. As of 2025, the Guangzhou Metro network operational mileage has reached 751.1 kilometers, and the total rail transit mileage in the Greater Bay Area exceeds 1,100 kilometers, forming an extensive and efficiently connected rail transit network.

Table 10-1 Components of Guangzhou Metro's Operation Management System

System Dimension	Core Content	Management Characteristics
Organizational Structure	Three-level management structure: Group-Line-Station	Unified command, hierarchical responsibility, professional collaboration
Institutional Standards	Comprehensive standard system covering operations, passenger transport, service,	Standardization, normalization, refinement

	safety, maintenance, etc.	
Technical Support	Smart systems including Tencent OS platform, big data, AI, IoT, etc.	Data-driven, intelligent decision-making, efficient execution
Performance Evaluation	Multi-dimensional assessment including safety, service, and efficiency indicators	Objective-oriented, continuous improvement

In terms of management system construction, Guangzhou Metro innovatively proposed the refined operation concept of "One Station, One Plan," requiring customized passenger service organization and service schemes tailored to the specific conditions of each station, including passenger flow characteristics, equipment configuration, and spatial layout. This concept fully demonstrates the forward-looking and adaptive nature of Guangzhou Metro's operation management, providing valuable experience for the networked operation of urban rail transit systems globally.

10.2 Intelligent Operation Platform System

The intelligent operation platform serves as the "brain" and "central nervous system" of modern metro operation management. Guangzhou Metro's innovative practices in this field are leading the industry's development direction. The Tencent OS intelligent operating system, spearheaded by Guangzhou Metro as the industry's first such system based on the industrial internet and Internet of Things (IoT), achieves deep integration and intelligent synergy among various specialized rail transit systems.

10.2.1 Intelligent Operation Control Platform

During the construction of Guangzhou Metro Line 10 and Line 12, the intelligent operation control platform independently developed by Panyu ICT equipped the lines with intelligent "station brains." This platform deeply integrates over 20 subsystems from different manufacturers and utilizes different protocols. Through data fusion and cross-equipment coordination, it enables intelligent assisted operation for scenarios such as one-click station startup/shutdown, video inspection, passenger flow early warning, operational dispatch, emergency plan activation, and maintenance assistance.

The platform adopts a three-layer architecture of "Data Layer - Platform Layer - Application Layer": the Data Layer is responsible for the collection and storage of multi-source data; the Platform Layer provides algorithm engines and data processing capabilities; the Application Layer offers intelligent functional modules for business scenarios. This open architecture not only addresses the pain points of traditional rail transit siloed architectures but also lays a solid foundation for subsequent system function iteration and upgrades.

10.2.2 Standardization of Intelligent Operation Platforms

Guangzhou Metro's intelligent initiatives extend beyond technological application to the level of industry standard leadership. The national standard "Intelligent Operation Platform for Urban Rail Transit," for which Guangzhou Metro Group serves as the primary drafting unit, has completed its project approval. This standard will fill the gap in foundational platform technology during the development of urban rail transit informatization and intellectualization.

This standard specifies the technical requirements for urban rail transit intelligent operation platforms, covering overall requirements and architecture, access, data, development, security, and operation maintenance. It aims to build more interconnected, intelligent, open, and collaborative urban rail transit intelligent operation platforms. The formulation of this standard will promote the deep integration of industrial internet, IoT, big data, and artificial intelligence with rail transit, construct a digitalized, networked, and intelligent smart rail transit system, and facilitate the transformation of rail transit from scale advantage to innovation advantage.

10.3 Passenger Service Organization and Service Management System

Passenger service organization and service are core aspects of metro operation management, directly impacting the passenger travel experience and the image of urban public transport. Guangzhou Metro has developed a scientific and comprehensive management system in this area and has made breakthrough progress in intelligent services.

10.3.1 Precise Passenger Flow Management Technology

Facing continuously growing passenger flow pressure, Guangzhou Metro has innovatively adopted advanced technologies such as "multi-modal integrated passenger flow prediction" and "dynamic flow restriction intelligent decision-making," achieving big data and AI-driven precise passenger flow management. These technologies provide core support for alleviating transport pressure on high-volume lines like Line 3, enabling comprehensive dynamic early warning, precise guidance, and capacity optimization across stations.

Since 2025, with the opening of Line 10 and Line 12, the accessibility of the Guangzhou Metro network has been further enhanced, highlighting the network synergy effect. Data shows that the new lines added 10 interchange stations, including Donghu, Xunfenggang, Dongxiaonan, and Wufeng, forming an efficient transfer system with existing lines and effectively balancing passenger flow distribution across the network. Against this backdrop, through precise passenger flow monitoring and data analysis, Guangzhou Metro has eliminated routine passenger flow control measures at 26 stations, reducing the number of stations with such controls by 41% compared to 2024.

Table 10-2 Guangzhou Metro Passenger Flow Control Levels and Response Measures

Control Level	Activation Conditions	Primary Measures	Application Scenario
Station-level Control	Platform passenger flow density exceeds warning threshold	Deactivate some access gates, adjust escalator directions	High passenger flow pressure at individual stations
Line-level Control	Station-level controls insufficient to alleviate passenger flow pressure	Coordinated passenger flow restrictions across multiple stations on the affected line	High passenger flow pressure across entire line

Network-level Control	Section passenger flow load factor exceeds warning threshold	Coordinated passenger flow restrictions on affected line and transfer lines	Large-scale passenger flow congestion

10.3.2 Application of Intelligent Service Robots

At the Guangzhou Baiyun Station Comprehensive Transport Hub, Guangzhou Metro has piloted the application of the "Wuyang" intelligent robot series. These robots cover five major functions: passenger service, luggage transport, patrol service, cleaning, and aerial patrols, forming a comprehensive intelligent service matrix.

•**Yang Xiaoying: A** customer service robot that breaks the spatial limitations of traditional service desks. Equipped with a 19-inch HD display screen, it integrates voice guidance and real-time voice interaction. It can guide passengers through security checks and civilized boarding, providing convenient services like voice inquiries and information queries.

•**Yang Xiaoyun: A** luggage robot specifically designed to solve the problem of passengers carrying large luggage. Using SLAM navigation and multi-sensor fusion technology, it can autonomously plan paths and intelligently avoid obstacles, allowing for easy travel.

•**Yang Xiaoxun: A** patrol service robot equipped with a LiDAR navigation system, enabling all-weather dynamic patrols of equipment, facilities, and passenger flow. Fitted with dual "visible light + thermal imaging" cameras and a smoke detection system, it can accurately identify abnormal situations such as water accumulation, smoke, and passenger falls, linking in real-time with the station control room.

•**Yang Xiaojie: A** cleaning robot integrating sweeping, mopping, and dust pushing functions, innovatively equipped with an AI intelligent inspection system. It can accurately identify types of ground stains and automatically match cleaning solutions, maintaining concourse cleanliness around the clock.

•**Yang Xiaofei: An** aerial patrol robot utilizing "low-altitude + 3D" technology, equipped with high-precision LiDAR and a thermal imaging gimbal. Applying 3D modeling and panoramic aerial photography, it achieves real-time video capture and dynamic monitoring of entrance/exit flood prevention, perfecting the metro's "ground-air integrated" emergency response system layout.

These robots not only function individually but also, through integration with the Guangzhou Metro smart platform, establish a collaborative linkage model of "intelligent discovery - precise guidance - efficient execution." Leveraging the powerful coordination capabilities of the metro smart platform, they utilize the management efficiency of the big data system platform to comprehensively enhance passenger service quality and efficiency.

10.3.3 Intelligent Ticketing and Passenger Service

Guangzhou Metro has widely applied millisecond-level facial recognition fare gate technology on Lines 10 and 12. Passengers can self-register their faces at intelligent customer service centers, activate the facial recognition fare gate function, and link it with the Guangzhou Metro APP ride code to achieve "code-face mixed use." The system's facial recognition latency is less than 0.3 seconds, truly achieving frictionless, non-stop rapid passage and significantly improving passenger entry and exit efficiency.

Simultaneously, Guangzhou Metro strictly adheres to the requirements of the "Measures for the Management of Urban Rail Transit Passenger Service Organization and Service," ensuring that lines operate for no less than 15 hours daily. Based on passenger flow demand, it formulates train operation plans, increases capacity during peak hours, and continuously improves the passenger service experience.

10.4 Intelligent Maintenance and Safety Assurance System

Intelligent maintenance and safety assurance form the cornerstone of metro operation management. Through technological innovation and management optimization, Guangzhou Metro has built a comprehensive, multi-layered maintenance and safety assurance system.

10.4.1 Fully Automatic Operation Technology

Guangzhou Metro Line 10 utilizes an FAO system of the highest automation grade (GoA4) for passenger service, achieving unattended operation throughout the entire process including train wake-up, self-test, operation, and dormancy. This represents a historic breakthrough for Guangzhou rail transit. This technology features dual modes for automatic and manual sleep/wake-up and supports functions like automatic depot entry/exit and automatic departure.

To ensure the safe operation of fully automatic trains, Panyu ICT conducted a comprehensive upgrade of the platform screen door system. By assigning independent numbers, each platform and each platform screen door has a unique "identity ID." This allows for rapid and accurate localization of specific platform screen doors during routine inspections, fault 排查, or remote real-time monitoring. The platform screen door system "partners" with the signaling system, enabling bidirectional, real-time exchange of door isolation status information and platform screen door status information, achieving precise linkage between platform screen doors and train doors, thereby enhancing train operation safety and efficiency.

10.4.2 Proactive Safety and Maintenance System

The National Engineering Research Center for System Safety and Operation Maintenance Assurance of Urban Rail Transit, led by Guangzhou Metro, is the only national strategic scientific and technological force in the domestic urban rail transit field established by a metro construction and operation enterprise. It has overcome core "bottleneck" technologies such as in-transit train monitoring and safety early warning, proactive maintenance of critical operation equipment and facilities, and major operational risk prevention and control.

In the operational safety capability enhancement project launched in 2025, Guangzhou Metro plans to systematically optimize and upgrade the safety operation capability of equipment and facilities on existing lines. This includes intelligently upgrading and enhancing security screening equipment on various lines; adding LiDAR early warning systems to 3 external overpass areas on Line 4; renovating poorly conditioned cross-passage doors in sections of Lines 1, 2, and 8; rectifying the security level protection of the Line 2 signaling ATS system and SCADA system;

adding intelligent access management functions for key network areas and contractor personnel management functions; and implementing capacity enhancement optimization renovations at 6 high-passenger-flow stations including Jiahe Wanggang. These measures will comprehensively enhance Guangzhou Metro's safety assurance and safety management capabilities, effectively addressing issues such as insufficient station capacity.

10.4.3 Emergency Management System

Guangzhou Metro has established a comprehensive emergency management system, developing detailed emergency plans for incidents such as fires, flooding, public security events, and public health emergencies. When an emergency occurs at a station, station staff must report to the traffic control department and conduct on-site according to the emergency plan, taking measures such as closing entrances/exits, evacuating passengers from the station, or closing the station if necessary.

Regarding emergency facilities, the "Yang Xiaoxun" and "Yang Xiaofei" from the "Wuyang" robot series form a "ground-air integrated" emergency monitoring network. They can accurately identify abnormal situations like water accumulation, smoke, and passenger falls, linking in real-time with the station control room, forming a new paradigm of human-machine collaborative service: "dynamic patrol - intelligent early warning - timely response," significantly improving emergency response speed.

10.5 Equipment and Facilities Maintenance Management System

Equipment and facilities maintenance is the material foundation for ensuring the safe and efficient operation of the metro. Guangzhou Metro has developed a full lifecycle management system in this area.

10.5.1 Intelligent Maintenance Platform

Guangzhou Metro integrates multiple functions such as equipment monitoring, fault early warning, maintenance dispatch, and spare parts management through its intelligent maintenance platform, achieving a shift from reactive repair to proactive prevention. Based on big data analysis, the platform can predict equipment failure trends, schedule maintenance plans in advance, effectively reduce equipment failure rates, and improve maintenance efficiency.

In the intelligent station systems of Lines 10 and 12, intelligent environmental control and equipment monitoring systems are deeply integrated with the comprehensive monitoring system, realizing the intelligent management of station equipment and facilities. This integrated management mode not only reduces maintenance costs but also improves system reliability and stability.

10.5.2 Full Lifecycle Management

Guangzhou Metro implements full lifecycle management for critical equipment. From equipment selection, installation and commissioning, operation and maintenance, to renewal and scrapping, each stage has clear technical standards and management specifications. Particularly regarding equipment renewal and renovation, Guangzhou Metro systematically upgrades aging equipment on existing lines, such as rectifying the security level protection of the Line 2 signaling ATS system and SCADA system, ensuring that equipment performance matches operational requirements.

10.6 Internationalization of Operation Management and Vocational Education Transformation

Transforming Guangzhou Metro's advanced operation management experience into international vocational education resources requires building a systematic knowledge transfer system and training curriculum.

10.6.1 Curriculum System Design

Based on the Guangzhou Metro operation management system, a "four-in-one" curriculum system integrating "Basic Theory - Technical Application - Case Practice - Management Thinking" can be constructed:

- **Basic Theory Module: Covers** fundamental theoretical knowledge including operation management concepts, organizational structure, institutional standards, and performance evaluation.

- **Technical Application Module: Provides** in-depth explanation of the principles and applications of key technologies such as the Suiteng OS platform, intelligent operation control, intelligent maintenance, and fully automatic operation.

- **Case Practice Module: Uses** typical Guangzhou Metro lines and stations as examples to analyze practical solutions for issues like passenger flow organization, emergency management, and service optimization.
- **Management Thinking Module: Cultivates** trainees' systems thinking, data analysis capabilities, and innovative awareness, enhancing comprehensive management qualities.

10.6.2 Training Platform Development

To enhance the effectiveness of vocational education, the following training platforms can be developed:

- **Intelligent Operation Simulation Platform: A** simulation platform developed based on the real Guangzhou Metro system, allowing trainees to experience the entire operation management process, including train operation organization, passenger flow management, and emergency response, in a virtual environment.
- **Intelligent Equipment Operation Platform: Provides** real equipment or high-fidelity models of facial recognition gates, intelligent robots, etc., enabling trainees to master the operation and maintenance skills for intelligent equipment.
- **Data Analysis Training Platform: Provides** desensitized operational data from Guangzhou Metro to train trainees' data analysis skills and foster data-based decision-making thinking.

10.6.3 International Adaptation Recommendations

When disseminating Guangzhou Metro's operation management experience internationally, adaptability to different countries and regions must be considered:

- **Technical Standard Adaptation: Extract** universally applicable technical principles and management concepts from Guangzhou's experience, adapting them to the technical standard systems of different countries.
- **Cultural Habit Integration: Consider** differences in passenger behavior habits and management models across cultural contexts to make targeted adjustments to operation management schemes.

•**Localized Implementation Strategy: Develop** practical localized implementation paths and phased goals according to the development stages and resource conditions of different regions.

10.7 Conclusion

Through years of operation management practice, Guangzhou Metro has formed a modern operation management system supported by intelligent platforms, based on standardized management, centered on precise services, and guaranteed by proactive safety. Its innovative practices in areas such as the Suzhou OS system, fully automatic operation, intelligent robots, and precise passenger flow management provide advanced solutions and a rich reference of experience for global urban rail transit operation management.

Looking ahead, with the advancement of the "14th Five-Year Plan" safety capability enhancement project and the formulation and implementation of the national "Intelligent Operation Platform for Urban Rail Transit" standard, Guangzhou Metro's operation management system will be further refined, providing safer, more efficient, and smarter rail transit services for the construction of the world-class city cluster in the Guangdong-Hong Kong-Macao Greater Bay Area. For international vocational education, systematically structuring, theorizing, and curricularizing these mature operation management experiences to cultivate new-type rail transit talents with modern operation management concepts and technical capabilities will be a key force in promoting the sustainable development of global rail transit.

Chapter 11: Vehicle System Maintenance and Overhaul

11.1 Overview of the Vehicle Maintenance and Overhaul System

As the core mobile assets of the urban rail transit system, the quality of metro vehicle maintenance and overhaul directly impacts operational safety, service quality, and economic efficiency. After more than two decades of development, Guangzhou Metro has established a modern vehicle maintenance and overhaul system guided by the principle of "Safety First, Prevention Foremost, Comprehensive Management." This system is based on planned maintenance, guided by condition-based maintenance, and supplemented by corrective maintenance, comprehensively covering key components such as vehicle bodies, bogies, traction systems, braking systems, auxiliary systems, network control systems, and passenger service facilities.

As of 2025, the operational mileage of the Guangzhou Metro network has reached 751.1 kilometers. The vehicle fleet includes multiple types such as Type A, Type B, Type L, and Regional Type D, forming a diversified vehicle portfolio adapted to the needs of different lines. Facing complex vehicle technologies and operational environments, Guangzhou Metro has innovatively proposed the refined maintenance strategy of "One Vehicle, One Plan," tailoring overhaul procedures and maintenance standards according to the technical characteristics and operational conditions of different vehicle models.

Table 11-1 Guangzhou Metro Vehicle Maintenance Levels and Content Classification

Maintenance Level	Cycle Range	Main Tasks	Implementation Location
Daily Inspection	Daily or every 2 days	Basic inspection of vehicle exterior, running gear, cab functions, etc.	Stabling Track
Fortnightly/Monthly Inspection	2 weeks - 1 month	Functional testing of key components, lubrication	Stabling Yard / Depot

		replenishment, filter cleaning	
Intermediate Overhaul	1-2 years	Partial component disassembly and repair, system function testing and calibration	Depot
Major Bogie Overhaul	5-6 years	Disassembly and repair of large components like bogies and traction motors	Depot
General Overhaul	10-13 years	Comprehensive disassembly, inspection, repair, and refurbishment of the entire vehicle	Vehicle Overhaul Workshop

The vehicle maintenance and overhaul system of Guangzhou Metro follows a development path of "standardization, normalization, informatization, and intelligentization." Building upon traditional planned maintenance, it has progressively introduced real-time monitoring-based predictive maintenance technologies, forming a closed-loop system for full lifecycle management. The establishment and implementation of this system have led to a continuous decrease in vehicle failure rates and a steady improvement in punctuality, offering valuable insights for global urban rail transit vehicle maintenance practices.

11.2 Maintenance System and Standardization Framework

Standardization serves as the technical foundation and management basis for vehicle maintenance work. While strictly adhering to national and industry standards, Guangzhou Metro actively participates in the formulation of industry standards and transforms its practical experience into internal technical specifications, forming a well-structured and comprehensive standardization framework.

11.2.1 Maintenance Regime and Procedures

Guangzhou Metro employs a maintenance regime that combines periodic maintenance with condition-based maintenance, defining different maintenance levels based on vehicle operating time or mileage. Specific maintenance levels include Daily Inspection, Fortnightly Inspection, Monthly Inspection, Three-Monthly Inspection,

Intermediate Overhaul, Major Bogie Overhaul, and General Overhaul, forming a multi-level maintenance system progressing from basic to comprehensive, and from superficial to in-depth checks.

Regarding the formulation of maintenance procedures, Guangzhou Metro has developed detailed maintenance process cards and work instructions tailored to the characteristics of different vehicle types. These documents clearly define the technical requirements, operational steps, quality standards, and safety precautions for various tasks. Taking the General Overhaul of the Guangfo Line B3-type vehicles as an example, the maintenance procedure specifies technical parameters and acceptance criteria for key steps such as battery capacity testing and charging/discharging maintenance processes.

11.2.2 Standardized Technical System

Guangzhou Metro actively participates in the development of industry standards, contributing as a main co-editing unit to the formulation of several standards, including the "Technical Specification for Operation of Urban Rail Transit Regional Express Vehicles" (T/CCTAS 237-2025). These standards provide unified technical specifications for the operation and maintenance of regional express vehicles, promoting the standardized development of industry technologies.

Internally, Guangzhou Metro has established a standardized system covering multiple dimensions such as maintenance procedures, process standards, quality acceptance, and safety management. Notably, before the introduction of new vehicle types like the green and intelligent urban rail vehicles on Line 12, Guangzhou Metro proactively conducted research on maintenance procedures and developed corresponding maintenance standards and training systems, ensuring a smooth transition for the operation and maintenance of the new models.

11.3 Maintenance Technologies for Key Components

The maintenance of key vehicle components is a core element in ensuring operational safety. Through long-term practice, Guangzhou Metro has developed a systematic fault analysis and repair technology system for the key components of different vehicle types.

11.3.1 Bogie System Maintenance Technology

As the vehicle's "running gear," the technical condition of the bogie directly impacts operational safety and stability. Building upon traditional bogie maintenance, Guangzhou Metro has developed specialized maintenance procedures for the new axle-box integrated frame-mounted direct-drive bogies used in the new green and intelligent urban rail vehicles.

By eliminating the traditional gearbox, this bogie type features a more compact structure, reduced overall bogie weight, and a significant decrease in unsprung mass, substantially enhancing bogie operational safety. Key inspection areas during maintenance include:

(1) **Axle-box integrated assembly: Inspection** of bearing clearance, seal condition, and lubrication status.

(2) **Direct-drive motor: Detection** of stator-rotor air gap, winding insulation resistance, and temperature sensor performance.

(3) **Suspension system: Inspection** of the fatigue state and geometric dimensions of components such as rubber joints and vertical links.

For the unique bogie structure of linear motor vehicles, Guangzhou Metro developed specialized tools and adjustment procedures for linear motor air gap measurement. By modifying the motor height adjustment method and improving the service life of rubber and metal joints, the occurrence of linear motor sagging faults has been effectively prevented.

The braking system is a critical safeguard for vehicle safety. Guangzhou Metro's vehicle braking systems have evolved technologically from air brakes to electromechanical brakes, with corresponding adjustments to maintenance strategies.

For traditional air brake systems, maintenance priorities include:

(1) **Brake caliper units: Inspection** of brake pad clearance, operational flexibility, and seal condition.

(2) **Air pipelines: Detection** of leakage, dryer performance, and filter blockage level.

(3) **Brake control units: Testing** of electro-pneumatic conversion performance and anti-skid control functions.

For the electromechanical braking system adopted by the green and intelligent urban rail vehicles on Line 12, compared to traditional air brakes, the response efficiency is improved by approximately 30%, with higher braking force accuracy and a smoother, more comfortable braking process. The maintenance focus for this system shifts to the inspection and calibration of motor actuators, transmission devices, and control algorithms, enabling rapid fault localization and intelligent diagnosis, thereby improving maintenance efficiency.

11.3.2 Traction System Maintenance Technology

The traction system serves as the "heart" of the vehicle, and its reliability directly affects operational performance. Guangzhou Metro has developed differentiated maintenance strategies tailored to the characteristics of traction systems across different vehicle models.

For power unit failures in L-type vehicle traction systems, disassembly analysis identified inadequate heat dissipation of the internal Intelligent Power Modules (IPMs) as the primary cause. Consequently, Guangzhou Metro established standards for IPM installation and application in power units, implementing measures such as cleaning heat sinks, replacing thermal grease, and promptly intercepting underperforming IPMs. These actions have significantly enhanced traction system reliability.

For the full silicon carbide converter permanent magnet traction system employed in Line 12's green intelligent rail vehicles, Guangzhou Metro has developed a predictive maintenance strategy based on condition monitoring. By real-time monitoring of converter output characteristics, permanent magnet motor temperature rise, and vibration data—integrated with the intelligent analytics capabilities of the Tencent OS platform—early warning of traction system faults and precise maintenance interventions are achieved.

11.3.3 Braking System Maintenance Technology

Table 11-2 Technical Features and Maintenance Priorities of Key Systems for Main Vehicle Types in

Vehicle Type	Traction System Features	Braking System Features	Special Maintenance Requirements
L-Type Linear Motor Vehicle	Linear induction motor drive	Air braking system	Linear motor gap adjustment and sagging prevention
B-Type Conventional Vehicle	Rotary motor with gear transmission	Air braking system	Traction motor bearing lubrication and gearbox inspection
A-Type Green Intelligent Rail Vehicle	Permanent magnet synchronous traction system	Electromechanical braking system	Full silicon carbide converter monitoring and permanent magnet motor maintenance
Regional D-Type Vehicle	High-power AC asynchronous traction	Composite braking system	High-speed operation adaptability adjustments and component reinforcement

11.4 Fault Diagnosis and Intelligent Maintenance Technology

With the development of IoT, big data, and artificial intelligence technologies, Guangzhou Metro has made significant progress in vehicle fault diagnosis and intelligent maintenance, establishing a smart maintenance system integrating data collection, intelligent analysis, and precise repair.

11.4.1 Bogie Temperature Monitoring Technology

Bogie temperature is a key parameter reflecting the vehicle's operational status. Guangzhou Metro has installed temperature sensors at critical locations on the vehicle bogie to monitor the temperature rise of components like bearings and gearboxes in real-time. The monitored data is transmitted wirelessly to a ground-based analysis system, which automatically triggers an alarm when temperatures exceed thresholds, guiding maintenance personnel to intervene promptly.

Guangzhou Metro participated in formulating the specification "Urban Rail Transit Metro Train Bogie Temperature Monitoring System Specification" (T/CQCBDS 0001-2025), which standardizes the specifications for urban rail train bogie temperature monitoring systems, enhancing their reliability and interoperability. This standard defines the complete set of technical requirements from temperature sensors, wiring harness devices, monitoring hosts, to the maintenance and analysis platform, providing a basis for the standardized design of such systems.

11.4.2 Intelligent Maintenance for Fully Automatic Operation Vehicles

The FAO system of the highest automation grade (GoA4) adopted on Line 10 presents new requirements for vehicle maintenance. The fully automatic operation vehicles are equipped with more comprehensive self-diagnostic systems capable of monitoring the status of key train components in real time and transmitting status data to the Suiteng OS platform.

After intelligent analysis of the data by the Tencent OS platform, the following functions can be achieved:

(1) Fault Prediction: Identifying component performance degradation trends and predicting potential faults based on historical data and operational parameters.

(2) Maintenance Decision-Making: Automatically generating maintenance recommendations and handling plans according to the nature and impact of faults.

(3) Resource Allocation: Automating train maintenance sequencing and work order dispatch to optimize maintenance resource allocation.

Additionally, Line 10 trains are equipped with integrated detection systems for monitoring the pantograph-catenary relationship and wheel-rail relationship, enabling them to "grasp" the real-time matching status based on actual operational conditions to ensure good pantograph-catenary and wheel-rail contact.

11.4.3 Big Data Analysis Technology

Guangzhou Metro utilizes big data technology to build a vehicle health management system. Through the mining and analysis of massive operational data, precise assessment of vehicle status and intelligent fault diagnosis are achieved. The system's main functions include:

(1) Performance Trend Analysis: Tracking changes in key component performance parameters and providing early warnings for performance degradation.

(2) Fault Correlation Analysis: Mining intrinsic relationships between different

faults to identify systemic risks.

(3) Maintenance Effectiveness Evaluation: Comparing parameter changes before and after maintenance to assess maintenance quality and identify areas for improvement.

Taking traction system power unit failure as an example, by analyzing historical temperature data and fault records of Intelligent Power Modules, a heat dissipation performance evaluation model was established, providing data support for optimizing maintenance strategies.

11.5 Green and Intelligent Vehicle Maintenance Innovation

The green and intelligent rail vehicles on Line 12 represent the latest development direction in vehicle technology, and their maintenance concepts and methods have been innovated and upgraded accordingly.

11.5.1 Green and Environmentally Friendly Maintenance Technologies

The green and intelligent rail vehicles incorporate multiple new energy-saving and environmentally friendly technologies, such as the full silicon carbide converter permanent magnet traction system. Compared to traditional asynchronous traction vehicles, the whole vehicle can achieve energy savings of over 15%. Correspondingly, Guangzhou Metro has introduced a green maintenance concept during the maintenance process, with specific measures including:

(1) **Application of Eco-friendly Materials: Using** naturally degradable floor coverings and employing electrostatic powder spraying technology for interior materials, achieving zero VOC emissions during the manufacturing process.

(2) **Waste Reduction and Resource Recovery: Classifying** and collecting waste oil, waste liquids, and solid waste generated during maintenance to improve resource reuse rates.

(3) **Energy-saving Operation Modes: Optimizing** maintenance processes, reducing equipment idle time, and lowering energy consumption during maintenance activities.

11.5.2 Predictive Maintenance Technology

The green and intelligent rail vehicles are equipped with a more comprehensive sensor network and intelligent diagnostic system, providing the technical foundation for predictive maintenance. The main characteristics of predictive maintenance include:

(1) Comprehensive Condition Perception: Real-time collection of operational parameters from various vehicle systems through multi-sensor fusion technology.

(2) Accurate Fault Diagnosis: Identifying abnormal equipment states and locating the root cause of faults based on artificial intelligence algorithms.

(3) Intelligent Maintenance Decision-Making: Scientifically scheduling maintenance timing and scope based on equipment health status and operational requirements.

Taking the electromechanical braking system as an example, the system can monitor the temperature of the friction pair in real time, predict the remaining life of the brake pads by combining historical data, and trigger replacement operations at the optimal time, thereby avoiding both over-maintenance and under-maintenance.

11.6 Typical Fault Analysis and Handling

Fault analysis is an important way to improve maintenance procedures and enhance maintenance levels. Guangzhou Metro has established a systematic fault collection, analysis, and improvement mechanism, continuously optimizing maintenance strategies through the analysis of typical cases.

11.6.1 Linear Motor Sagging Fault Analysis

The L-type vehicles on Guangzhou Metro Lines 4, 5, and 6 use linear motors, where the distance from the bottom of the linear motor to the reaction plate has strict requirements. Linear motor sagging can seriously affect operational safety.

Through fault analysis, the main causes identified for linear motor sagging include:

(1) Pin Shaft Wear: Long-term vibration leads to increased clearance between the pin shaft and the mounting base.

(2) Failure of Motor Height Adjustment Device: Wear or corrosion of the

adjustment thread pair leads to insufficient locking force.

(3) Rubber Joint Aging: Degradation of rubber material properties leads to stiffness changes and displacement accumulation.

(4) Abnormal Vertical Link: Component fatigue or loosening of connections causes positional changes.

In response to the above causes, Guangzhou Metro formulated corresponding inspection procedures and improvement measures, such as changing the motor height adjustment method, improving the service life of rubber joints and metal joints, and strictly controlling the production quality and factory inspection items of various components.

11.7 Vocational Education Transformation Pathways

Transforming Guangzhou Metro's advanced vehicle maintenance experience into international vocational education resources requires building a systematic knowledge transfer system and training curriculum.

11.7.1 Curriculum System Design

Based on the Guangzhou Metro vehicle maintenance system, a "four-in-one" curriculum system integrating "Basic Theory - Technical Application - Fault Analysis - Innovative Practice" can be constructed:

(1) Basic Theory Module: Covers basic theoretical knowledge such as vehicle structure, working principles, and maintenance theory, explaining the technical characteristics of different vehicle types combined with Guangzhou Metro's vehicle portfolio.

(2) Technical Application Module: Provides in-depth explanation of the principles and key operational points of key technologies such as bogie maintenance, traction system maintenance, and braking system testing, incorporating Guangzhou Metro's standardized operating procedures.

(3) Fault Analysis Module: Uses typical faults like linear motor sagging and battery power loss as cases to cultivate trainees' fault analysis thinking and problem-solving abilities.

(4) Innovative Practice Module: Introduces maintenance concepts and methods for new technology vehicles like green and intelligent rail vehicles, cultivating trainees' innovative awareness and adaptability.

11.7.2 Training Platform Development

To enhance the effectiveness of vocational education, the following training platforms can be developed:

(1) Key Component Disassembly/Assembly Training Platform: Provides real components or high-fidelity models of bogies, traction motors, brake units, etc., enabling trainees to master standardized disassembly/assembly processes and adjustment methods.

(2) Fault Diagnosis Simulation Platform: Builds a simulation system based on real fault cases from Guangzhou Metro to train trainees' fault troubleshooting and diagnostic skills.

(3) Intelligent Maintenance Operation Platform: Provides a simplified version or simulation system of the Suiteng OS platform, familiarizing trainees with the operation methods and data analysis techniques of intelligent maintenance systems.

11.7.3 International Adaptation Recommendations

When disseminating Guangzhou Metro's vehicle maintenance experience internationally, adaptability to different countries and regions must be considered:

(1) Technical Standard Adaptation: Extract universally applicable technical principles and management concepts from Guangzhou's experience, adapting them to the technical standard systems of different countries.

(2) Resource Condition Adaptation: Design tiered training content and differentiated implementation plans considering the equipment conditions and technical levels of different regions.

(3) Cultural Environment Adaptation: Respect local cultural habits and work styles, making appropriate adjustments to training methods and communication styles.

11.8 Conclusion

Through years of vehicle maintenance practice, Guangzhou Metro has established a modern vehicle maintenance system based on a standardized system, centered on preventive maintenance, directed towards intelligent maintenance, and aimed at full lifecycle management. Its innovative practices in areas such as linear motor maintenance, permanent magnet traction system overhaul, bogie temperature monitoring, and fully automatic operation vehicle maintenance provide advanced solutions and a rich reference of experience for global urban rail transit vehicle maintenance.

Looking ahead, with the batch deployment of green and intelligent rail vehicles and the continuous upgrade of the Suzhou OS platform, Guangzhou Metro's vehicle maintenance system will further develop towards digitalization, intelligence, and greening, providing safer, more reliable, and more efficient vehicle support for the construction of the world-class city cluster in the Guangdong-Hong Kong-Macao Greater Bay Area. For international vocational education, systematically structuring, theorizing, and curricularizing these mature vehicle maintenance experiences to cultivate new-type rail transit talents with modern maintenance concepts and technical capabilities will be a key force in promoting the sustainable development of global rail transit.

Chapter 12:Signaling and Communication System Operation and Maintenance

12.1 Overview of Signaling and Communication Systems

As the "nerve center" and "sensory system" of urban rail transit, signaling and communication systems undertake critical functions including ensuring operational safety, improving transport efficiency, and supporting passenger services. After over two decades of development, Guangzhou Metro has established a modern signaling and communication system characterized by automated train command, automated train operation, and integrated information transmission. This system comprehensively covers key subsystems such as Automatic Train Control (ATC), Automatic Train Supervision (ATS), Computer-based Interlocking (CI), Data Communication System (DCS), Transmission System, Wireless Communication System, Administration Telephone System, and Video Surveillance System.

The signaling systems encompass various advanced technologies including Communication-Based Train Control (CBTC) and Fully Automatic Operation (FAO), forming a diversified technological portfolio adapted to the needs of different lines. Facing complex operational environments and technical challenges, Guangzhou Metro has innovatively proposed a system operation and maintenance philosophy of "Intelligent Perception, Precise Control, Efficient Transmission, and Smart Maintenance," developing tailor-made maintenance strategies and technical standards according to the technical characteristics and operational requirements of different systems.

Table 12-1 Main Subsystems and Functions of Guangzhou Metro Signaling and Communication Systems

System Category	Main Subsystem	Core Functions	Maintenance Characteristics

	Automatic Train Control (ATC)	Train overspeed protection, automatic operation, precise stopping	High safety and real-time requirements
Signaling System	Automatic Train Supervision (ATS)	Train operation monitoring, schedule adjustment, operation statistics	High reliability, human-machine interaction
	Computer-based Interlocking (CI)	Control of switches, signals, track circuits	Fault-safety principle
	Data Communication System (DCS)	Bidirectional continuous train-to-ground communication	High bandwidth, low latency
	Transmission System	Transmission of voice, data, video information	Networked, large capacity
Communication System	Wireless Communication Syst	Train-to-ground voice communication, mobile TV	Comprehensive coverage, high interference resistance
	Administration Telephone System	Internal administrative communication, inter-station calls	High reliability, easy scalability
	Video Surveillance System	Real-time monitoring of stations and trains	High definition, intelligent analysis

The operation and maintenance of Guangzhou Metro's signaling and communication systems adhere to the development path of "prevention-first, combining prevention and control, technology empowerment, and refined management." Building upon traditional planned maintenance, predictive maintenance technologies based on real-time monitoring have been gradually introduced, forming a closed-loop system for full lifecycle management. The establishment and implementation of this system have led to a continuous decrease in signaling system failure rates and a steady improvement in communication service quality, providing valuable experience for the operation and maintenance of signaling and communication systems in global urban rail transit.

12.2 Signaling System Operation and Maintenance Technology

The signaling system is the core guarantee for the safe and efficient operation of the metro, and its maintenance quality directly impacts operational safety and transport efficiency. Through long-term practice, Guangzhou Metro has developed a systematic maintenance system and innovative solutions for signaling systems with different technical characteristics.

12.2.1 Fully Automatic Operation System Maintenance Technology

Guangzhou Metro Lines 10 and 12 have implemented the highest automation grade (GoA4) FAO system for passenger service, achieving unattended operation throughout the entire process including train wake-up, self-test, operation, and dormancy. This represents a historic breakthrough for Guangzhou rail transit. The maintenance of this technology differs significantly from traditional signaling systems, presenting new requirements for maintenance work.

Fully automatic operation vehicles are equipped with more comprehensive self-diagnostic systems capable of monitoring the status of key train components in real time and transmitting status data to the Suiteng OS platform. To ensure the safe operation of fully automatic trains, Panyu ICT conducted a comprehensive upgrade of the platform screen door system. By assigning independent numbers, each platform and each platform screen door was given a unique "identity ID," enabling rapid and accurate localization of specific platform screen doors during routine inspections, fault troubleshooting, or remote real-time monitoring.

The platform screen door system "partners" with the signaling system, enabling bidirectional, real-time exchange of door isolation status information and platform screen door status information. This achieves precise linkage between platform screen doors and train doors, enhancing both train operation safety and efficiency. This technological innovation not only addresses key safety hazards in the FAO system but also provides convenience for system maintenance.

12.2.2 Automatic Train Supervision System High Availability Technology

The ATS system is the "command center" of the signaling system, and its reliability directly affects the operational order of trains across the entire line. In maintaining the ATS system, Guangzhou Metro innovatively adopted Linux dual-network transformation and high-availability cluster technology, significantly improving system reliability.

- **Physical Layer Hard Isolation: Each** server was configured with 4 network cards – two for the signaling control network (connecting to trackside equipment) and two for the dispatch command network (connecting to the control center). This is equivalent to expanding the original single road into two independent highways!
- **Logical Layer Dynamic Switching: Through** bonding technology, two network cards of the same type were bundled into a virtual network port. If the primary network card fails, the standby card takes over instantly (0 seconds), with no impact on upper-layer services.
- **High Availability Cluster Architecture: Uses** 1 master node to handle real-time train control commands (e.g., departure, speed adjustment), and 3 standby nodes for distributed processing of massive status monitoring (position, speed, signals). If the master node fails, a new master is automatically elected within 10 seconds (much faster than maintenance staff can reach the equipment room!). During network fluctuations, the cluster performs self-checking and seamless traffic switching – not even a pop-up will appear on the train dispatch screen!

Violent testing verified that during simulated physical power loss of the master node during morning peak hours, switching succeeded in 9 seconds, with control transferred to a standby node. When the primary signaling control network card was intentionally cut, the backup card took over in 0 seconds, maintaining uninterrupted communication with trackside equipment. When both network failure and master database process termination were simulated simultaneously, the cluster self-healed within 15 seconds, sending automatic alerts directly to the duty officer's phone. This technology compresses failure switchover time from hours to seconds, significantly enhancing signaling system availability.

12.2.3 Signaling System Renewal and Transformation Technology

According to the latest tender plan released by Guangzhou Metro Group, signaling system renewal and transformation projects will commence for Guangzhou Rail Transit Lines 2, 3, 8, 5, and the APM Line. This renovation focuses on lines with longer operational years and aging equipment. For instance, the initial phases of Lines 2 and 8 have been in operation for over 23 years since opening in 2002, while Lines 3, 5, and the APM Line have also been operational for at least nearly 15 years.

The tender plan specifies that the renovation content covers system-wide upgrades including control centers, mainline, depots, onboard equipment, and interfaces. The aim is to enhance operational safety redundancy and operational efficiency through technological iteration. Specific renovation projects include:

(1) **Line 2 and Line 8 Renovation: Tender** announcements are planned for August 2025, involving comprehensive upgrades of the existing signaling systems.

(2) **Line 3 Renovation + Line 5 and APM Line Renovation: Design** tenders are launched simultaneously. For the APM Line, as a medium-capacity rubber-tired system, the renovation scope also includes coordinated updates for vehicles, communications, integrated supervision, and the platform screen door system.

These renovation projects will effectively address the declining reliability caused by equipment aging, while simultaneously improving system performance and maintenance efficiency through technological upgrades.

Table 12-2 Comparison of Key Technologies in Guangzhou Metro Signaling System Operation and Maintenance

Technology Category	Traditional Technology	Innovative Technology	Maintenance Benefits
System Architecture	Single Network, Single	Linux Dual Network +	Fault switchover

	Node	High Availability Cluster	reduced from hours to seconds
Fault Handling	Manual On-site Handling	Remote Diagnosis + Automatic Switching	70% improvement in maintenance efficiency
Status Monitoring	Periodic Inspections	Real-time Monitoring + Predictive Warning	Increased proportion of preventive maintenance
Interface Management	Independent Maintenance	Full System Linkage	Enhanced system coordination

12.3 Communication System Operation and Maintenance Technology

The communication system serves as the "information highway" for the metro, providing fundamental support for operations management, passenger services, emergency command, and other functions. Guangzhou Metro has achieved significant innovations in the field of communication system operation and maintenance, particularly in 5G communication networks and cloud-edge collaborative architecture.

12.3.1 5G Communication Network Operation and Maintenance Technology

Recently, with the completion of the 5G communication network upgrade project for Line 5 and the Northern Extension of Line 3, the Guangzhou Metro network has achieved continuous 5G signal coverage across all tunnel sections and platforms. This milestone marks the official entry of the entire Guangzhou Metro network into the 5G era. Beyond installing 5G signals in newly constructed lines, this comprehensive upgrade project has also substantially "enhanced" mobile signals across the entire existing network. Passengers can now experience smoother network services whether waiting on platforms or traveling inside moving trains.

The signal upgrade work was primarily concentrated in platform areas and tunnel sections. The confined space of the metro tunnels presented significant challenges: equipment and facilities were highly integrated, coordination was complex, precision requirements were extremely high, construction was difficult, and safety risks were elevated. Furthermore, persistent high passenger flow added considerable complexity to the upgrade project. To avoid disrupting normal operations on existing lines, construction was confined to a 2.5-hour "golden window" each night after train services ended. At the peak of the upgrade activities, over 800 personnel were organized to work simultaneously across 12 different lines.

To comprehensively enhance construction efficiency, Guangzhou Metro and Guangdong Tower standardized safety control procedures by integrating management models. They organized various operational departments to establish construction specifications, categorizing 1,096 special scenarios into nine conventional coverage methods, including adding installations, fixing clamps, and jumper connections. An efficient "sectional slicing" construction model was also developed, where work was delivered sequentially based on train schedules - completing and handing over one tunnel section at a time. This approach successfully facilitated the installation of 1.7 million clamps, laying of 2.7 million meters of cable, and commissioning of 8,300 devices within the tunnels of existing lines. Consequently, continuous 5G signal coverage across the entire network was achieved four months ahead of the original schedule.

Technological innovation was crucial for the rapid 5G network upgrade. To this end, Guangzhou Metro and Guangdong Tower collaborated closely. On one hand, they organized multi-party technical demonstrations to develop customized, integrated products combining fiber optic boxes, POI (Point of Interface), and combiners, which substantially reduced both the quantity of device installations required and the overall construction timeline. On the other hand, they innovated with digital tools such as smart helmets and body cameras, enabling remote supervision, real-time feedback, and rapid response, thereby comprehensively strengthening quality and safety management during the tunnel construction and upgrades. Following the achievement of continuous 5G coverage, the average perceived downlink rate for trains operating in tunnels increased from 70-80 Mbps to 400-600 Mbps, significantly enhancing passenger internet browsing speeds.

12.3.2 Cloud-Edge Collaborative Network Architecture Operation

Guangzhou Metro Design & Research Institute Co., Ltd. has filed a patent for a "Cloud-Edge Collaborative Network Architecture System for Urban Rail Transit

Networks." This system comprises at least one central station and multiple non-central stations. The central stations, which are transfer stations, line starting stations, or line terminal stations, are deployed with edge cloud nodes and configured with primary cloud resource clusters, first standby cloud resource clusters, and second standby cloud resource clusters.

Non-central stations, while not equipped with edge cloud nodes, connect to central stations via primary and backup links and are similarly configured with primary and standby cloud resource clusters. The Network Operation Control Center (NOCC) and Line Operation Control Centers (OCCs) are deployed at the central stations. Edge computing management center nodes are also deployed within the NOCC and OCCs, responsible for handling the scheduling of network data traffic.

This patent integrates the operational characteristics of the urban rail transit industry, the networking methods of communication networks, and the latest development trends in cloud-edge collaboration technology. It designs a network architecture solution suitable for truly achieving cloud-edge collaboration within urban rail transit networks, thereby significantly enhancing overall operational reliability. This architecture provides flexible, efficient resource scheduling and fault tolerance capabilities for communication system operation and maintenance.

12.3.3 Communication System Renewal and Transformation Management

On September 30, 2025, the bid winning results for the design service project of the communication system renewal and transformation for Guangzhou Metro's existing lines were announced. Guangzhou Metro Design & Research Institute Co., Ltd. won the bid with a total contract value of 17.20241 million RMB. This project signifies that Guangzhou Metro's communication system renewal and transformation work has entered the substantive implementation phase, aiming to enhance communication system performance and operational efficiency through technological iteration.

Regarding 5G-Advanced technology application, China Telecom Guangzhou Branch has completed Guangzhou's first batch of metro 5G-A activation projects. Targeting the metro's "pulse-type" passenger flow characteristics, a second-level load monitoring standard highly matched with the metro environment was established. This enables more sensitive detection of areas where user experience is impacted by instantaneous traffic surges when trains arrive, providing a precise basis for timely resource allocation. This capacity expansion was implemented autonomously. Teams from China Telecom Guangzhou Branch coordinated closely, carrying out site access coordination, board installation, data configuration, and operational testing methodically. Within just one month, they completed the 5G-A activation and bandwidth doubling for key stations. Traffic volume increased substantially by 44%, and the perceived rate improved by 35% to 323 Mbps, significantly alleviating previous traffic congestion issues and providing a smoother passenger internet experience.

12.4 Intelligent Operation Platform and Technology

With the development of IoT, big data, and artificial intelligence technologies, Guangzhou Metro has made significant progress in the intelligent operation and maintenance of signaling and communication systems, establishing a smart maintenance system integrating data collection, intelligent analysis, and precise repair.

12.4.1 Large Model-based Intelligent Operation Technology

A patent named "Intelligent Operation and Maintenance Method and Device for Rail Transit Communication System Based on Large Model" provides an innovative solution for intelligent operation of signaling and communication systems. The method is applied on an intelligent computing platform, which is obtained by training a general large model based on professional data corresponding to the rail transit communication system.

The method includes: acquiring system real-time data corresponding to the rail transit communication system collected by sensing devices; determining intelligent operation instructions corresponding to the rail transit communication system based on the real-time data; sending the intelligent operation instructions to execution devices. The instructions direct the execution devices to execute the intelligent operation plan corresponding to the instructions for the rail transit communication system. This technology uses a general large model to train the intelligent computing platform, enabling intelligent operation of the rail transit communication system through sensing devices, the intelligent computing platform, and execution devices, thereby improving operation efficiency and accuracy while reducing labor costs.

12.4.2 Intelligent Operation Control Platform

During the construction of Guangzhou Metro Lines 10 and 12, the intelligent operation control platform independently developed by Panyu ICT equipped the lines with intelligent "station brains." This platform deeply integrates over 20 subsystems from different manufacturers and utilizes different protocols. Through data fusion and cross-equipment coordination, it enables intelligent assisted operation for scenarios such as one-click station startup/shutdown, video inspection, passenger flow early warning, operational dispatch, emergency plan activation, and maintenance assistance.

The platform adopts a three-layer architecture of "Data Layer - Platform Layer - Application Layer": the Data Layer is responsible for the collection and storage of multi-source data; the Platform Layer provides algorithm engines and data processing capabilities; the Application Layer offers intelligent functional modules for business scenarios. This open architecture not only addresses the pain points of traditional rail transit siloed architectures but also lays a solid foundation for subsequent system function iteration and upgrades.

12.4.3 Platform Screen Door Intelligent Control System

Guangzhou Metro Group Co., Ltd. has applied for a patent named "A Platform Screen Door Control System and Method," which discloses an embedded-based platform screen door control system and method. The control system includes: a signal input module, a logic module, and a signal output module. The input end of the signal input module is connected to external signals; the input end of the logic module is connected to the output end of the signal input module and the output end of the signal output module; the output end of the logic module is connected to the input end of the signal input module and the input end of the signal output module; the output end of the signal output module is connected to the platform screen doors.

The signal input module includes two mutually redundant input channels; the signal output module includes two mutually redundant output channels. Redundant detection of external signals is achieved through the signal input module; redundant control of the platform screen doors is achieved through the signal output module; the connection relationships between the signal input module, logic module, and signal output module form a closed-loop control. Combined with redundant detection and control, this improves system reliability. This technological innovation enhances the reliability of the platform screen door system as a key interface device between signaling and trains.

12.5 Typical Fault Analysis and Handling

Fault analysis is an important way to improve operation procedures and enhance maintenance levels. Guangzhou Metro has established a systematic fault collection, analysis, and improvement mechanism, continuously optimizing operation strategies through the analysis of typical cases.

12.5.1 Network Switching Fault Analysis

Before the Linux dual-network transformation of the Guangzhou Metro ATS system, network switching failures were a major factor affecting system reliability. Through fault analysis, the main causes of network switching failure were identified as:

(1) **Inadequate Heartbeat Detection Mechanism: Traditional** heartbeat detection was inaccurate in judging transient network interruption scenarios.

(2) **Unreasonable Switching Threshold Settings: Overly** sensitive settings caused frequent switching, while overly lenient settings led to switching delays.

(3) **Redundant Path Quality Differences: Inconsistent** physical path quality between primary and backup network cards resulted in incorrect status judgments.

(4) **Driver Compatibility Issues: Mismatches** between network card drivers and operating system versions caused abnormal status reporting.

In response to these causes, Guangzhou Metro implemented corresponding optimization measures in the Linux dual-network transformation, such as improving the heartbeat detection algorithm, optimizing switching threshold settings, standardizing primary/backup path quality standards, and strictly enforcing driver compatibility testing. These measures significantly enhanced the reliability of network switching.

12.5.2 5G Network Capacity Insufficiency Analysis

Before the Guangzhou Metro 5G network upgrade, insufficient network capacity at some stations during peak hours was a major issue affecting passenger experience. Through systematic investigation, the fault causes were mainly concentrated in the following areas:

(1) "Pulse-type" Passenger Flow Characteristic: Instantaneous high-density user access when trains arrived caused network overload.

(2) Unreasonable Cell Coverage Range: Excessively large coverage area for a single cell led to an overloaded number of users.

(3) Mismatched Resource Allocation Strategy: Fixed resource allocation could not adapt to dynamic service demands.

(4) Equipment Performance Limitations: Outdated equipment with insufficient processing capacity limited throughput.

Based on the analysis results, Guangzhou Metro took targeted measures in the 5G upgrade project, including establishing a second-level load monitoring standard, optimizing cell coverage ranges, introducing dynamic resource allocation strategies, and upgrading outdated equipment. These measures effectively resolved the network capacity insufficiency issue during peak hours.

12.6 Vocational Education Transformation Pathways

Transforming Guangzhou Metro's advanced experience in signaling and communication system operation and maintenance into international vocational education resources requires building a systematic knowledge transfer system and training curriculum.

12.6.1 Curriculum System Design

Based on Guangzhou Metro's signaling and communication system operation and maintenance framework, a "four-in-one" curriculum system integrating "Basic Theory - Technical Application - Fault Analysis - Innovative Practice" can be constructed:

(1) **Basic Theory Module: Covers** fundamental theoretical knowledge including signaling and communication system principles, network architectures, and interface protocols. Explains the technical characteristics of different system types in conjunction with Guangzhou Metro's system portfolio.

(2) **Technical Application Module: Provides** in-depth explanation of the principles and key operational points of core technologies such as ATS system operation and maintenance, 5G network optimization, cloud-edge collaborative management, and platform screen door control. Integrates Guangzhou Metro's standardized operating procedures.

(3) **Fault Analysis Module: Uses** typical faults like network switching failures and capacity insufficiency as case studies to cultivate trainees' fault analysis thinking and problem-solving abilities.

(4) **Innovative Practice Module: Introduces** operation and maintenance concepts and methods for new technology systems such as FAO systems and large model-based intelligent maintenance, fostering trainees' innovation awareness and adaptability.

12.6.2 Training Platform Development

To enhance the effectiveness of vocational education, the following training platforms can be developed:

(1) **Signaling System Simulation Platform: Provides** simulation environments for key systems like ATS and CI, enabling trainees to master system operation, parameter configuration, and fault handling procedures.

(2) **Communication Network Testing Platform: Provides** 5G network testing tools and analysis software to train trainees' network optimization and fault diagnosis skills.

(3) **Intelligent Operation Platform: Provides** a simplified version or

simulation system of the Tencent OS platform, familiarizing trainees with the operation methods and data analysis techniques of intelligent operation systems.

12.6.3 International Adaptation Recommendations

When disseminating Guangzhou Metro's signaling and communication system operation experience internationally, adaptability to different countries and regions must be considered:

(1) **Technical Standard Adaptation: Extract** universally applicable technical principles and management concepts from Guangzhou's experience, adapting them to the technical standard systems of different countries.

(2) **Resource Condition Adaptation: Design** tiered training content and differentiated implementation plans considering the equipment conditions and technical levels of different regions.

(3) **Cultural Environment Adaptation: Respect** local cultural habits and work styles, making appropriate adjustments to training methods and communication styles.

12.7 Conclusion

Through years of practice in signaling and communication system operation and maintenance, Guangzhou Metro has established a modern operation system based on a standardized framework, centered on preventive maintenance, directed towards intelligent operation, and aimed at full lifecycle management. Its innovative practices in areas such as FAO system operation, ATS high-availability technology, 5G network upgrades, cloud-edge collaborative architecture, and large model-based intelligent operation provide advanced solutions and a rich reference of experience for the operation and maintenance of global urban rail transit signaling and communication systems.

Looking ahead, with the continuous upgrade of the Tencent OS platform and the formulation and implementation of the national "Intelligent Operation Platform for Urban Rail Transit" standard, Guangzhou Metro's signaling and communication system operation framework will further develop towards digitalization, intelligence, and cloud-network integration. This will provide safer, more reliable, and efficient communication and signaling support for the construction of the world-class city cluster in the Guangdong-Hong Kong-Macao Greater Bay Area. For international vocational education, systematically structuring, theorizing, and curricularizing these mature operation experiences to cultivate new-type rail transit talents with modern operation concepts and technical capabilities will be a key force in promoting the sustainable development of global rail transit.

Chapter 13:Power Supply and Electromechanical System Operation and Maintenance

13.1 Power Supply System Architecture and Equipment Characteristics

The Guangzhou Metro power supply system employs a hierarchical distributed structure composed of main substations, traction substations, step-down substations, and distribution networks, forming a comprehensive electrical energy supply system that covers the entire network. Serving as the "heart" of the metro system, the power supply system is responsible for converting high-voltage electricity from the municipal grid into appropriate specifications for metro trains and various equipment. Main substations undertake the task of transforming voltage from 110 kV to 33 kV, transmitting power to station substations via cable lines. Traction substations further reduce the voltage from 33 kV to 1500 V DC to provide traction power for metro trains, while step-down substations lower the voltage to 380 V/220 V to meet the electricity demands of station equipment such as lighting, ventilation, communications, and signaling systems.

The Kengkou Main Substation of Guangzhou Metro is a typical representative of this power supply system. As the first main substation of Guangzhou Metro, it has been responsible for ensuring the power supply for Line 1 since its commissioning in 1997. In line with the requirements of the "13th Five-Year Plan" for line construction, this substation will also undertake the power supply task for the under-construction Line 10. After more than two decades of continuous operation, all electrical equipment in the substation has entered its "aging phase," reaching the scheduled period for major overhaul and replacement. Issues such as equipment aging and operational maintenance difficulties have become increasingly prominent. The technical performance and reliability of core components, particularly the 110 kV GIS equipment, main transformers, and 33 kV GIS, directly impact the operational safety and efficiency of the entire metro network.

Table 13-1 Main Equipment and Functions of Guangzhou Metro Power Supply System

Equipment Type	Main Function	Technical Characteristics	Typical Location
110 kV GIS Equipment	Interface with city grid, high-voltage power reception and distribution	Fully enclosed combined electrical apparatus, high reliability, compact footprint	Main substation
Main Transformer	110 kV to 33 kV voltage transformation	Oil-immersed or dry-type transformer, typically 20-40 MVA capacity	Main substation
33 kV GIS Equipment	Medium-voltage power distribution and protection	Modular design, easy maintenance	Main substation, Traction substation
DC Traction Unit	AC to DC conversion, providing traction power for trains	Combination of rectifier transformer and silicon rectifier, 1500 V DC output	Traction substation
Low-voltage Distribution Cabinet	380 V/220 V power distribution	Drawer-type switchgear, multiple circuit outputs	Step-down substation at each station

The architectural design of Guangzhou Metro's power supply system embodies the principles of redundant backup and zonal power supply. Critical lines such as Line 1 are equipped with two independent power source points. When one power source fails, the other can assume the power supply task for the entire line, ensuring uninterrupted operation. This design philosophy was validated during the 2023 power supply equipment failure incident on Line 1. Although the equipment failure caused operational disruptions in some sections, the system's redundant design successfully prevented a larger-scale operational paralysis.

13.2 Operation Challenges and Aging Equipment Management

With the rapid expansion of the Guangzhou Metro network and the aging of early line equipment, the operation and maintenance of power supply and electromechanical systems face unprecedented challenges. As of 2024, the Guangzhou Metro network mileage ranks third nationally, while its passenger flow intensity ranks first. This leads to rapid equipment wear and extremely high reliability requirements. Facilities like the Kengkou Main Substation, which has been in continuous operation for twenty-seven years, are particularly affected. All electrical equipment in such substations has entered an "aging phase," with issues like equipment deterioration and increasing maintenance difficulties becoming increasingly prominent.

These aging problems mainly manifest as: decreased insulation performance, increased mechanical operating resistance, poor contact in control circuits, and supply difficulties due to discontinued spare parts. For instance, GIS equipment commissioned in the 1990s commonly exhibits aging and hardening of seals within its operating mechanisms, leading to altered operating characteristics and increased failure risk. Simultaneously, early relay protection devices utilize outdated technology with limited fault diagnosis capabilities, struggling to meet the modern metro system's requirements for power supply reliability.

Confronting these challenges, Guangzhou Metro innovatively proposed a "live-line renovation" construction scheme. The core of this strategy is to phase out aging equipment while ensuring normal operation of existing lines. During the Kengkou Main Substation renovation project, technicians achieved effective isolation between construction areas and operational electrical equipment through "zonal handover of the operating main substation" and the use of 110 kV "double-break isolation earthing switches." This method ensured that two GIS sections could be connected without requiring simultaneous power outages, thus maintaining continuous GIS operation. This approach guaranteed both construction safety and the stable operation of the original equipment.

Table 13-2 Primary Issues and Countermeasures for Aging Power Supply Equipment in Guangzhou Metro

Aging Issue	Impact on System	Countermeasures	Application Case
Insulation Performance Degradation	Increased short-circuit risk; reduced reliability	Online monitoring & early warning; regular insulation testing	Cable Insulation Monitoring at Kengkou Station
Mechanical Component Wear	Deterioration of operational characteristics; increased failure rate	Mechanical characteristic monitoring; regular lubrication maintenance	GIS Mechanism Retrofit
Control Circuit Aging	Slow response speed; risk of malfunction	Upgrade to microprocessor-based protection devices	Relay Protection System Upgrade
Spare Parts Discontinuation	Extended maintenance cycles; increased costs	Complete equipment replacement; standardized design	Full Station Electrical Equipment Replacement
Outdated Technical Standards	Poor compatibility with newly added equipment	Interface standardization; technical upgrade	Intelligent Retrofit of Kengkou Station

13.3 Research and Application of Intelligent Maintenance Technologies

Confronted with the characteristics of megacity metro networks—"large scale, extensive line distribution, numerous hubs, and high operational load intensity"—Guangzhou Metro actively promotes the intelligent transformation of power supply and electromechanical system maintenance. This shifts the paradigm from the traditional "manual experience-based" maintenance model to a data-driven intelligent maintenance mode.

13.3.1 Online Monitoring and Early Warning Technologies

In the technical renovation of the Kengkou Main Substation, Guangzhou Metro innovatively employed online monitoring and digital twin technologies to establish an integrated system centered on an online monitoring system, a condition assessment system, and an equipment information management system. By installing online monitoring devices on primary power supply equipment, this system acquires real-time online data of the equipment's core state variables, enabling monitoring, early warning, analysis, and diagnostic functions. Specifically, the system can monitor key parameters such as SF6 gas density in GIS equipment, circuit breaker mechanical characteristics, cable joint temperature, and surge arrester leakage current. It assesses the health status of the equipment and provides early warnings through intelligent algorithms.

The application of the online monitoring system has enabled Guangzhou Metro to achieve automated inspection, automatic early warnings, and automated report generation for various equipment within the main substation. This system not only reduces the workload of maintenance personnel but, more importantly, facilitates the timely detection of potential equipment defects, preventing failures. For instance, by continuously monitoring the temperature trend of cable joints, the system can issue warnings before overheating occurs due to poor contact, alerting maintenance personnel to address the issue promptly and thereby avoiding more widespread power supply failures.

13.3.2 Application of Intelligent Inspection Robots

In 2025, Guangzhou Metro, in collaboration with Southern Power Grid Technology, successfully piloted its self-developed, lightweight indoor intelligent inspection robot in the 400V low-voltage room of the step-down substation at Kengkou Station on Metro Line 1. This robot represents the power industry's first ten-thousand-yuan-level intelligent indoor inspection product, characterized by low cost and high reliability. It is equipped with autonomous obstacle avoidance functionality, allows for flexible task planning via the intelligent inspection system, and supports remote control, monitoring, and autonomous target observation. Capable of autonomous return and charging, it enables round-the-clock online inspection.

The application of the intelligent inspection robot effectively addresses challenges inherent in traditional manual inspection modes, such as low efficiency, high labor intensity, and insufficient real-time capability. The robot can automatically identify equipment status—including switchgear indicator lights, protection circuit links, and meter readings—as well as assess environmental conditions. This significantly alleviates the workload of maintenance personnel and facilitates the intelligent upgrade of substation operation and maintenance. In the pilot application at Guangzhou Metro, the robot replaced approximately 40% of the manual inspection workload and increased the inspection frequency by threefold, markedly enhancing the completeness and timeliness of equipment condition data.

13.3.3 Equipment Maintenance Refined Management Information System

Leveraging years of operational and management experience, and aligning with the requirements of a refined management system and the characteristics of rail transit facilities and equipment, Guangzhou Metro introduced technologies such as QR code tags and handheld terminal devices to develop the Equipment Maintenance Refined Management Information System (LMIS). The LMIS is built upon the philosophy of refined management and the concept of full life-cycle management for equipment and assets. It aims to achieve automatic field equipment identification and automatic dispatch of equipment-related tasks, thereby enhancing field personnel capabilities in positioning and safety protection, and supporting the operational processes of equipment refined management.

The core of the LMIS is its management framework centered on the entire life cycle of equipment maintenance, which involves a comprehensive review and detailed standardization of maintenance procedures, work steps, and maintenance models across various technical disciplines. The system strengthens supervision over the maintenance process and control over work quality, improves dispatch management capabilities, enhances technical support for frontline production, and establishes a refined equipment maintenance management system. This ensures standardized management of equipment maintenance operations, integrated planning, guided operations, monitored execution, post-event traceability, and data analyzability.

Since its launch in April 2014, the maintenance module has been deployed across all disciplines (25 maintenance specialties) on more than four Guangzhou Metro lines. By September 2017, it was extended to cover all lines of the network. The system is used by tens of thousands of users, including all of Guangzhou Metro's maintenance staff, contracted personnel, and technicians, making it a vital technical platform supporting the operation and maintenance of the metro's power supply and electromechanical equipment.

13.3.4 CPS-based Intelligent Maintenance System

Guangzhou Metro has established an intelligent maintenance system incorporating Cyber-Physical Systems (CPS) characteristics, enabling online equipment condition monitoring, fault diagnosis, and dynamic inventory adjustment. This system constructs precise mapping relationships between power supply equipment and their digital models through three key technologies: cyber-physical integration, real-time perception, and dynamic interaction.

The CPS-based intelligent maintenance system primarily consists of the following components:

(1) Intelligent Perception Layer: Deploys various sensors and intelligent inspection robots at critical nodes of power supply equipment to collect real-time equipment status data.

(2) Network Transmission Layer: Utilizes converged wired and wireless communication networks to achieve reliable data transmission.

(3) Data Platform Layer: Integrates, stores, and manages multi-source heterogeneous data.

(4) Intelligent Application Layer: Implements equipment condition assessment, fault prediction, and maintenance decision support using artificial intelligence algorithms.

The implementation of this technical system has facilitated Guangzhou Metro's transition from "periodic maintenance" to "predictive maintenance" for power supply equipment, significantly enhancing equipment reliability while reducing operational costs.

Table 13-3 Comparison of Intelligent Maintenance Technology Application Effectiveness in Guangzhou Metro

Technical Type	**Baseline Status (Pre-Application)**	**Outcomes (Post-Application)**	**Typical Case**
Online Monitoring Technology	Manual inspection; discontinuous data	Real-time monitoring; automatic early warning	Online Monitoring System at Kengkou Station
Inspection Robots	Low-frequency manual inspection; high labor intensity	Round-the-clock automated inspection; improved efficiency	400V Low-Voltage Room at Kengkou Station
Digital Twin Technology	2D drawings; isolated information	3D visualization; holistic perception	Digital Twin Model of Main Substation
Refined Management Information System	Paper-based records; fragmented processes	Electronic workflows; whole-process control	LMIS Deployment Across Entire Net
CPS-based Intelligent Maintenance System	Reactive maintenance; scheduled repairs	Predictive maintenance; condition-based repairs	Intelligent Maintenance of Power Supply Equipment

13.4 Emergency Management and Fault Handling Mechanisms

The emergency management of Guangzhou Metro's power supply system adheres to the principles of "safety first, prevention as priority, rapid response, and hierarchical handling," establishing a comprehensive mechanism for fault early warning, emergency response, and incident management. This mechanism was fully demonstrated during the power supply equipment failure incident that occurred during the morning peak hour on Line 1 on May 11, 2023.

13.4.1 Fault Emergency Response Process

On the morning of May 11, 2023, a power supply equipment failure occurred on Guangzhou Metro Line 1, causing delays in train services in both directions from

Xilang to Guangzhou East Railway Station. Following the incident, Guangzhou Metro promptly activated its emergency response mechanism:

(1) At 07:41, initial delay information was released via the official Weibo account, informing passengers of slightly extended train intervals on Line 1 from Xilang towards Guangzhou East Railway Station.

(2) At 08:07, the cause was further clarified as a power supply equipment failure, and passengers were advised to avoid the affected sections when planning their journeys.

(3) At 08:24, emergency bus bridging services were activated for the section from Fangcun to Dongshankou on Line 1. Passengers were recommended to use alternative transfer stations such as Huangsha, Chen Clan Academy, Gongyuanqian, Yangji, and Dongshankou for detours.

This emergency response process highlighted the characteristics of information transparency, rapid coordination, and multi-modal transport collaboration by Guangzhou Metro following a power supply failure. Through timely information dissemination and the implementation of emergency bus bridging services, the impact of the failure on passenger travel was minimized.

13.4.2 Rapid Power Restoration Strategy

After the power supply equipment failure occurred, Guangzhou Metro's technical team quickly located the fault point, analyzed the cause, and implemented effective isolation and restoration measures. According to reports, multiple citizens reported hearing a "loud bang" and witnessing a bright light between the stations before and after Chen Clan Academy. This indicated that the fault likely involved an electrical failure such as a short circuit or arc flash.

The power supply zoning and redundancy backup strategies considered in Guangzhou Metro's design played a significant role in this situation. Through fault isolation and power supply path switching, the technical staff successfully confined the impact to a local section, ensuring normal operation in other sections. By 08:57, Guangzhou Metro's official Weibo updated that the power supply equipment fault on Line 1 had been resolved and train intervals were gradually returning to normal.

It is worth mentioning that Guangzhou Metro also provided comprehensive follow-up service measures after the fault was handled, notifying passengers that they could handle related ticket procedures at any metro station (except the APM Line) within seven days with their held tickets. If passengers needed to obtain a 'Dear Passengers' letter, they could download it from the official Guangzhou Metro website, WeChat account, or APP, or contact station staff to obtain a copy. This comprehensive after-service system reflects Guangzhou Metro's service philosophy of "Always with You."

Table 13-4 Emergency Response Mechanism for Guangzhou Metro's Power Supply System

Emergency Phase	Key Measures	Responsible Department	Communication Mechanism
Prevention Stage	Equipment monitoring, risk assessment, contingency plan development	Equipment Maintenance Department, Safety Management Department	Internal notifications, early warning dissemination
Preparation Stage	Emergency resource stockpiling, personnel training, drills	Human Resources Department, Materials Management Department	Training notifications, drill feedback
Response Stage	Fault isolation, bus bridging, information release	Dispatch Center, Station Management, Publicity Department	Passenger notifications, media communication
Recovery Stage	Equipment restoration, gradual resumption of normal operations	Equipment Maintenance Department, Operations Dispatch Department	Recovery announcements, service notices
Post-Incident Stage	Fault analysis, improvement measures, service compensation	Technical Department, Customer Service Department, Safety Management Department	Incident reporting, rectification notices

13.5 International Vocational Education and Training System

To cultivate international technical talents meeting the requirements for intelligent maintenance of power supply and electromechanical systems, Guangzhou Metro has established a systematic vocational education and training system covering four levels: fundamental theory, technical skills, practical application, and international perspective.

13.5.1 Training System Framework

Guangzhou Metro's vocational education and training system is competency-based, designed according to the principles of "categorization by level, progressive learning, and integration of theory and practice." For personnel involved in power supply and electromechanical system operation and maintenance, the training content primarily includes:

- **Fundamental Theory Module:** Covers specialized foundational knowledge such as electrical fundamentals, power system analysis, relay protection principles, and motors and drives.
- **Specialized Technology Module:** Offers specialized tracks like high-voltage electrical equipment, relay protection, DC power supplies, and low-voltage power distribution tailored to different position requirements.
- **Intelligent Maintenance Module:** Includes content on online monitoring technologies, fault diagnosis methods, predictive maintenance strategies, and other intelligent O&M topics.
- **Safety Management Module:** Addresses knowledge of electrical safety regulations, emergency incident response, and the use of safety tools and equipment.
- **International Standards Module:** Introduces international standards from bodies like the International Electrotechnical Commission (IEC) and the Institute of Electrical and Electronics Engineers (IEEE), cultivating a global perspective among trainees.

This training system is implemented through various formats including classroom teaching, virtual simulation, hands-on training, and field internships, ensuring trainees can translate theoretical knowledge into practical working capabilities.

13.5.2 Specialized Skills Training for Intelligent Maintenance

In response to the development of intelligent maintenance technologies for the power supply system, Guangzhou Metro has specifically strengthened specialized skills training for intelligent maintenance within its vocational education program. Training content includes:

- **Application of Online Monitoring Devices:** Learning the principles, installation, and debugging methods of various sensors.

- **Data Analysis and Processing:** Mastering basic methods for data acquisition, processing, and analysis.

- **Equipment Condition Assessment:** Learning methods for assessing equipment health status based on monitoring data.

- **Predictive Maintenance Strategies:** Understanding how to develop maintenance plans based on equipment condition predictions.

Human-Robot Collaborative Inspection: Mastering the processes and methods for working collaboratively with intelligent inspection robots.

Taking the indoor lightweight intelligent inspection robot as an example, Guangzhou Metro conducted related operation and maintenance training alongside its pilot application. Technical personnel needed to learn the robot's basic principles, operational procedures, routine maintenance, and simple fault handling to ensure they could fully utilize the robot to improve inspection efficiency and quality.

13.5.3 International Curriculum Development and Teaching Resource Construction

To promote the international vocational education for power supply and electromechanical system operation and maintenance technologies, Guangzhou Metro collaborated with several internationally renowned enterprises and educational institutions to develop a series of international courses and teaching resources:

- **International Curriculum System:** Introduced internationally advanced standards and technologies, combined with local needs, to develop a curriculum system suited for the international market.

- **Bilingual Teaching Resources:** Compiled English-language training materials and produced bilingual teaching videos to meet international training demands.

- **Virtual Simulation Platform:** Developed a virtual simulation platform for the power supply system, enabling virtual training for practical skills like equipment disassembly/assembly, debugging, and fault handling.

- **Training Base Construction:** Built comprehensive training bases covering main substations, traction substations, and step-down substations, replicating real-site environments.

- **Teaching Faculty Development:** Cultivated a team of "dual-qualified" instructors possessing both theoretical teaching ability and field experience.

Through this comprehensive vocational education and training system, Guangzhou Metro has cultivated a large number of technical and skilled talents for power supply and electromechanical system O&M, not only meeting its own development needs but also providing a reference for talent cultivation in other city metro systems.

13.5.4 International Cooperation Model Integrating Industry, Academia, and Research

Guangzhou Metro actively promotes an international cooperation model integrating industry, academia, and research, establishing long-term cooperative relationships with numerous domestic and foreign universities, research institutions, and well-known enterprises. For instance, Guangzhou Metro participated in organizing the "Digital-Intelligent Interconnection Empowering High-Quality Development of Rail Transit" academic conference, inviting several renowned academicians in the rail transit field, representatives from universities, research institutes, and corporate executives to discuss and exchange ideas on industry hotspots such as key rail transit technologies, green and low-carbon development, and digital-intelligent transformation.

This industry-academia-research integrated international cooperation model provides advanced technical support and a resource-sharing platform for vocational education:

Technical Exchange: Keeps abreast of the latest international technology trends through international conferences, technical forums, etc.

- **Cooperative Research:** Collaborates with domestic and foreign universities and research institutions to conduct research on intelligent maintenance technologies for power supply systems.

- **Faculty Exchange:** Arranges for instructors to undertake exchange visits to advanced metro companies or training institutions abroad to enhance teaching standards.
- **Student Exchange:** Establishes student exchange mechanisms with foreign vocational colleges to broaden students' international horizons.
- **Standard Alignment:** Promotes the alignment of vocational qualifications with international standards, enhancing the international recognition of talent cultivation.

Through these international collaborations, Guangzhou Metro continuously absorbs advanced international experience to refine its own vocational education system, while also contributing Chinese solutions to international vocational education, achieving a transition from "bringing in" to "going global."

The development journey of Guangzhou Metro's power supply and electromechanical system operation and maintenance technologies provides valuable experience for global urban rail transit. From the early renewal and renovation of equipment, to the comprehensive application of intelligent maintenance technologies, and the construction of a systematic vocational education system, Guangzhou Metro has charted a sustainable development path from technological catch-up to innovation leadership. These experiences hold significant reference value for cities worldwide that are constructing or expanding their urban rail transit systems, and also provide rich teaching resources and practical cases for international vocational education in the rail transit field.

Chapter 14: Intelligent Stations and Smart Operation and Maintenance Systems

14.1 Intelligent Station System Architecture and Platform Design

Guangzhou Metro's intelligent station development centers around the Tencent OS (Rail Transit Intelligent Operating System), establishing an industry-pioneering new-generation rail transit operating system based on the Industrial Internet of Things (IIoT) and Internet of Things (IoT). Since its initial deployment at Canton Tower Station in 2019, this system has undergone iterative upgrades from version 1.0 to 2.0. It is now widely implemented across multiple lines of Guangzhou Metro, enabling intelligent perception, data interaction, intelligent analysis, and coordinated operation of various station equipment systems.

14.1.1 Evolution and Development of the Tencent OS System

Tencent OS is a rail transit operating system jointly developed by Guangzhou Metro and Tencent. Its development history reflects Guangzhou Metro's continuous innovation in the field of intelligent systems. In 2019, Tencent OS 1.0 was officially launched at the Canton Tower Smart Metro Demonstration Station. This initial version realized four core functions—intelligent security, intelligent passenger flow guidance, intelligent station management, and comprehensive information display—established the platform's foundational architecture, and met the application standards for urban rail transit line-level deployment.

In 2021, Guangzhou Metro further released Tencent OS 2.0, deploying it for demonstration operations on Lines 18 and 22. Version 2.0 focused on building five core platforms: the IoT Platform, Strategy Engine Platform, Big Data Platform, Algorithm Platform, and Open Platform. This version is characterized by standardization, componentization, low barriers to entry, and open architecture.

SuiTeng OS has obtained significant certifications, including the "Safety Integrity Level (SIL) 2 Certification" and the "Information System Security Level Protection (Level 3) Certification," alongside numerous invention patents and software copyrights. These achievements signify the initial formation of a complete and systematic set of scientific and technological outcomes. The system redefines the design philosophy of traditional industrial control systems by fostering comprehensive open sharing. This approach attracts more industry partners to participate in application and plugin development, thereby enriching the operating system's ecosystem.

Table 14-1 Comparative Analysis of Functional Characteristics Between Tencent OS Versions

Functional Module / Core Capability	**SuiTeng OS 1.0**	**SuiTeng OS 2.0**	**Application Lines / Deployment Scope**
IoT Connectivity Capability	Supported integration of over 20 equipment subsystems	Expanded to support integration with over 20 specialized professional systems, encompassing more than 20,000 individual devices	Canton Tower Station (1.0); Lines 18 & 22 (2.0)
Data Platform Capability	Provided foundational data collection and processing functionalities	Enhanced with a robust Big Data Platform and a dedicated Algorithm Platform, enabling advanced data analytics and intelligent processing	Phased rollout and implementation across the entire metro network
Intelligent Linkage Control	Enabled basic automated responses for predefined scenarios	Significantly advanced via a Strategy Engine Platform, incorporating over 400 analytical and control models for	Primarily demonstrated on Lines 18 and 22

		complex, adaptive automation	
Open Ecosystem	Featured limited openness for external development	Established a comprehensive open-sharing platform, actively encouraging third-party application and plugin development to foster ecosystem growth	Targeted for future planned lines and ongoing ecosystem expansion
Security Management	Incorporated fundamental security protection measures	Implemented a multi-layered security framework encompassing network security, data security, IoT device security, and algorithm security	Applied as a standard across the entire metro network

14.1.2 Intelligent Station System Architecture Design

The intelligent station system of Guangzhou Metro adopts a layered architecture, sequentially comprising the Perception Layer, Network Layer, Platform Layer, and Application Layer from bottom to top. This architectural design facilitates intelligent perception, data interaction, intelligent analysis, and coordinated operation of various station equipment systems, providing "coordinated and consistent" operational support for station services.

Perception Layer: At this foundational level, Guangzhou Metro has deployed a vast array of IoT sensors, intelligent cameras, smart robots, and other devices to collect real-time information on the station environment, equipment status, and passenger flow data. Taking Guangzhou Baiyun Station as an example, the station has centrally deployed intelligent service robots covering five major functions: passenger service, luggage transport, patrol service, cleaning, and aerial inspection. These include robots named "Yang Xiao Ying" (Welcoming Lamb), "Yang Xiao Yun" (Transporting Lamb), "Yang Xiao Xun" (Patrolling Lamb), "Yang Xiao Jie" (Cleaning Lamb), and "Yang Xiao Fei" (Flying Lamb). These robots constitute a comprehensive station operation perception network, providing the intelligent station with all-round data acquisition capabilities.

Network Layer: Guangzhou Metro employs a converged wired and wireless network architecture at this layer to ensure high-speed and reliable data transmission. Specifically designed to meet the data transmission demands during peak-hour large passenger flow scenarios, the Network Layer incorporates redundant backup systems and intelligent load balancing mechanisms to guarantee the real-time performance of critical data.

Platform Layer: The five core platforms of SuiTeng OS 2.0 – the IoT Platform, Strategy Engine Platform, Big Data Platform, Algorithm Platform, and Open Platform – provide the core technical support for the intelligent station. The IoT Platform enables unified access and management of various devices. The Strategy Engine Platform supports intelligent linkage for multiple business scenarios. The Big Data Platform provides data storage, processing, and analysis capabilities. The Algorithm Platform integrates various AI algorithms to support intelligent analysis and decision-making. The Open Platform attracts ecosystem partners to participate collaboratively in application development.

Application Layer: At the top layer, Guangzhou Metro's intelligent station realizes a series of smart functions, including intelligent security, intelligent passenger flow guidance, intelligent station management, and comprehensive information display. These applications are developed for specific station operation scenarios based on the capabilities provided by the Platform Layer, significantly enhancing station operational efficiency and service quality.

14.1.3 Standardized Construction of Intelligent Stations

Guangzhou Metro actively participates in the standardization of intelligent stations and led the development of the national standard "Intelligent Operation Platform for Urban Rail Transit." This standard specifies the technical content for urban rail transit intelligent operation platforms concerning general requirements and architecture, access, data, development, security, and operation maintenance. It is applicable to the design, research and development, production, and application of urban rail transit intelligent operation platforms.

The main content of the standard covers seven key aspects:

- **General Requirements: Specifies** the design principles, construction planning, and business requirements for the urban rail transit intelligent operation platform
- **Overall Architecture: Defines** the overall architecture of the platform, including comprehensive planning from the Data Layer, Platform Layer, and Application Layer, along with the content contained within each level.
- **Access Requirements: Stipulates** the technical requirements for platform access in three areas: IoT access, data access, and application access.
- **Data Requirements:** Specifies requirements for the platform's master data management, metadata management, data standard management, data quality management, data asset management, and data service management.
- **Application Development Requirements: Standardizes** requirements for the platform's development architecture, development principles, and development forms.
- **Security Requirements:** Regulates requirements for the platform's network security, data security, IoT security, and algorithm security.

Operation and Maintenance (O&M) Requirements: Standardizes aspects including O&M monitoring principles, O&M monitoring objects, routine maintenance, fault management, and change and release management.

The formulation of this national standard fills the gap in foundational platform technical standards during the development of urban rail transit informatization and intelligentization. It guides the intelligent development of urban rail transit and enhances the intelligence level of urban rail transit construction and operation management. Furthermore, this standard lays the foundation for the international promotion of Guangzhou Metro's intelligent station technology.

14.2 Core Application Systems of the Intelligent Station

Guangzhou Metro's intelligent station application systems, underpinned by the SuiTeng OS, have achieved a comprehensive intelligent transformation of station operations. These systems cover multiple business domains including passenger flow management, station affairs management, and security & emergency response, significantly enhancing operational efficiency and service quality.

14.2.1 Intelligent Passenger Flow Management System

The Intelligent Passenger Flow Management System is a core application of Guangzhou Metro's intelligent stations. It enables scenario-based coordinated control by real-time perception of passenger flow density in different areas of platforms and trains, analyzing the distribution of passengers, and automatically triggering key service devices such as mobile apps, station broadcasts, electronic guide screens, and escalators. This guides passengers to board trains in carriages with lower crowd density.

The system collects real-time passenger flow data through video analytics cameras and sensors deployed at key locations within the station. Combined with artificial intelligence algorithms, it analyzes and predicts passenger flow status. When the system detects that passenger density in a certain area exceeds a predefined threshold, it automatically initiates pre-set passenger flow control protocols. For instance, it can guide passengers via electronic guide screens to suggest alternative routes, adjust the operational direction of escalators, and issue guidance broadcasts, thereby balancing passenger distribution throughout the station and preventing congestion and potential safety hazards.

In the demonstration application on Guangzhou Metro Line 18, the Intelligent Passenger Flow Management System realized panoramic operational dispatch and intelligent passenger flow guidance, substantially improving the station's passenger handling capacity and the overall passenger experience. The system can also predict short-term passenger flow trends based on historical data and real-time conditions, providing decision-making support for station operation and management.

14.2.2 Intelligent Station Management System

The Intelligent Station Management System automates and intellectualizes daily station operations. One of its most typical applications is the "One-Click" Station Open/Close function. This system revolutionizes the traditional station service mode. It optimizes the previous process, which required station staff to physically go to each station entrance daily to manually open and close them, now enabling the opening or closing of the entire station with a single command. This greatly enhances station operational efficiency and service quality.

The Intelligent Station Management System also implements scenario-based mode control for functions like video-based patrol inspection and the one-click activation and full-process tracking of station emergency plans. It facilitates mobile station management, thereby reducing operational management costs. Leveraging AI technologies such as intelligent voice recognition and knowledge graphs, the system enhances the level of office automation and improves staff productivity.

Furthermore, escalators in the intelligent demonstration stations are equipped with an intelligent operation and maintenance system. This system enables real-time monitoring of escalator operating status, provides immediate fault alarms, and offers health assessment and diagnostics along with maintenance recommendations. This intelligent approach to equipment management not only improves device reliability but also reduces the workload associated with manual inspections.

14.2.3 Intelligent Security and Emergency Command System

The Intelligent Security System achieves comprehensive coverage of video surveillance throughout stations and tunnel sections, enabling rapid identification and early warning of anomalous behaviors. It facilitates unified management of security screening equipment, centralized image interpretation, and task dispatch, and can trigger one-click alarm activation when necessary. Leveraging artificial intelligence algorithms, the system automatically detects abnormal situations such as passenger falls, left-behind items, and crowd gatherings, promptly notifying station staff for intervention.

Regarding emergency command, the system provides specialized solutions for scenarios like flood prevention and control in rail transit. Station operation managers can monitor critical data in real-time, including waterlogging locations, water levels, and the distribution of emergency supplies. Once water levels reach predefined warning thresholds, the station's flood prevention emergency plan can be activated with a single command. This triggers the automatic dissemination of warning information and enables rapid coordination with station systems like broadcasts, escalators, and fare gates to facilitate orderly passenger evacuation.

In developing its intelligent stations, Guangzhou Metro has also introduced the aerial inspection robot "Yang Xiao Fei" (Flying Lamb). Utilizing "low-altitude + three-dimensional" technology and equipped with high-precision LiDAR and a thermal imaging gimbal, this robot employs 3D modeling and panoramic aerial photography. It performs functions such as real-time video capture and dynamic flood monitoring at station entrances/exits, thereby enhancing the metro's integrated "ground-air" emergency response framework. This combined ground and aerial intelligent security architecture significantly strengthens the overall safety assurance capability of the stations.

14.3 Intelligent Maintenance System and Technology Application

Guangzhou Metro's intelligent maintenance system is fundamentally data-driven. Leveraging advanced technologies such as the Internet of Things (IoT), big data, and artificial intelligence (AI), it achieves full life cycle management and predictive maintenance for equipment and facilities. This represents a significant shift from the traditional "preventive maintenance" paradigm towards a "predictive maintenance" model, substantially enhancing maintenance efficiency and equipment reliability.

14.3.1 Cloud-Edge Collaborative Intelligent Maintenance Architecture

In the domain of signaling system maintenance, Guangzhou Metro has developed and deployed an intelligent maintenance system for urban rail transit signaling based on a cloud-edge collaborative computing architecture. This system employs a four-layer technical architecture, comprising the Application Layer, Data Layer, Access Layer, and Perception Layer. It provides detailed specifications for both cloud and line-side equipment configurations. Aligned with the practical application requirements of the signaling intelligent maintenance system, it proposes corresponding solutions across five key areas: signaling onboard equipment maintenance, emergency fault handling, equipment room inspection and patrols, preventive maintenance, and production organizational structure.

The cloud-edge collaborative architecture fully leverages the respective advantages of the cloud and the edge. The cloud offers powerful computing and storage capabilities, suitable for large-scale data analysis and complex algorithm execution. The edge, being close to the data source, enables real-time data processing and rapid response. In Guangzhou Metro's implementation, the collection and processing of fundamental signaling system data are completed at the edge, ensuring rapid response for business operations with high real-time requirements. Conversely, big data analytics and intelligent decision-making are performed in the cloud, utilizing its robust computational power for deep data mining and model training.

This architecture has significantly improved maintenance efficiency and reduced operational costs in areas such as signaling equipment repair, inspection, and emergency duty. Particularly for emergency fault handling, the system can quickly locate fault points and provide handling recommendations, greatly shortening fault resolution time.

14.3.2 Full Life Cycle Equipment Health Management

The trains on Guangzhou Metro Line 10 utilize an intelligent maintenance system that implements intelligent maintenance throughout the entire train life cycle. This system monitors the status of key train components in real-time and instantly transmits the status data to Guangzhou Metro's rail transit intelligent operating system – the SuiTeng OS platform. After intelligently analyzing the data, the platform identifies potential issues and provides early warnings. It also automates train maintenance scheduling and work order dispatch, providing strong technical support for the safe and smooth operation of the trains.

The "Jia Yi Wei" (JiaYi Maintenance) intelligent maintenance system, developed by PDT Science & Technology Co., Ltd. and piloted on the second phases of

Guangzhou Metro Lines 13 and 14, deeply utilizes big data, IoT, and cloud computing technologies. It performs 7x24-hour "precise condition monitoring" of equipment and facilities, enabling proactive fault prediction and accurate condition assessment. This thoroughly innovates the traditional maintenance mode, transitioning from reactive repair to precise "condition-based maintenance." While ensuring safe and worry-free equipment operation, it empowers full life cycle management, achieving a dual leap in maintenance efficiency and cost-effectiveness.

Full life cycle equipment health management not only focuses on the real-time status of equipment but also integrates data from the entire process—from design and manufacturing through operation and maintenance, all the way to decommissioning—building a comprehensive equipment health profile. Through the analysis of this data, the system can evaluate the remaining useful life of equipment, optimize maintenance strategies, and reduce the total cost of ownership across the entire life cycle.

14.3.3 Functions and Applications of the Intelligent Maintenance Platform

The intelligent maintenance platform of Guangzhou Metro has achieved innovative applications across multiple specialized systems, enhancing operational safety, efficiency, and cost-effectiveness.

- **Vehicle Maintenance: On** Line 10 trains, an integrated detection system has been installed to monitor the pantograph-catenary interaction and wheel-rail relationship. This system continuously "grasps" their real-time matching status based on actual operating conditions. This capability is crucial for ensuring optimal pantograph-catenary current collection and proper wheel-rail mechanical interaction, thereby providing concrete and effective technical safeguards for train operational safety.

- **Platform Screen Door Systems: PDT** Science & Technology has applied its self-developed, compatible Platform Screen Door (PSD) Door Control Unit (DCU) to address challenges such as the difficulty in replacing components from different PSD brands and procuring imported spare parts. This solution effectively reduces operation and maintenance costs while simultaneously enhancing the safety and maintenance convenience of metro operations.

• **Energy Management: An** intelligent Environmental Control and Energy Saving System, based on hybrid models and AI algorithms, has been implemented. This system utilizes digital twin and deep learning technologies to perform real-time optimization of the Environmental Control System (ECS) within metro stations. It deeply explores energy-saving potential, leading to reduced operational energy consumption. By monitoring real-time environmental parameters inside and outside the station and integrating passenger flow data, the system dynamically adjusts the operational status of ventilation and air conditioning equipment. This ensures passenger comfort is maintained while minimizing energy usage to the greatest extent possible.

Furthermore, the intelligent maintenance platform integrates several core functions that create a complete, closed-loop maintenance management process:

• **Automated Work Order Dispatch: Generates** maintenance work orders automatically based on equipment condition data and analytical results.

• **Dynamic Resource Allocation:** Intelligently assigns tasks to appropriate maintenance personnel according to the urgency and complexity of the work order.

• **Real-time Process Monitoring: Tracks** the progress of maintenance activities in real-time.

Maintenance Quality Evaluation: Assesses the quality and effectiveness of completed maintenance work.

Maintenance personnel receive work orders, access historical equipment data, and review maintenance plans via mobile terminals. This streamlined process significantly improves both the efficiency and quality of maintenance operations.

Table 14-2 Core Technologies of Guangzhou Metro's Intelligent Maintenance System

Technology Type	Functional Characteristics	Application Scenario	Implementation Outcome

Cloud-Edge Collaborative Computing	Centralized cloud analytics and decision-making combined with real-time edge control and processing.	Intelligent maintenance of signaling systems.	Significantly improves maintenance efficiency and reduces operational costs.
Big Data Analytics	Integration and fusion of multi-source data; deep mining and correlation analysis for insights.	Equipment condition assessment and performance prediction.	Enables predictive maintenance, minimizing unplanned downtime and failures.
Digital Twin	Creates a high-fidelity virtual mapping of physical assets, supporting real-time interaction and simulation.	Optimization of Environmental Control Systems (ECS).	Uncovers deep energy-saving potential through system modeling and optimization.
Artificial Intelligence Algorithms	Capable of intelligent pattern recognition, predictive forecasting, and automated early warning.	Automated fault diagnosis and intelligent incident handling.	Enhances fault resolution speed and operational decision-making efficiency.
Internet of Things (IoT) Technology	Provides comprehensive situational awareness and continuous, real-time monitoring of equipment status.	Pervasive equipment condition monitoring across the network.	Achieves 7x24 "precise condition monitoring" for uninterrupted asset health oversight.

14.4 Intelligent Service Robot Technology and Application

In its intelligent station development, Guangzhou Metro has deployed intelligent service robots on a large scale to enhance station operational efficiency and service quality. In 2025, Guangzhou Metro initiated a large-scale pilot application of intelligent robots at the modern comprehensive transportation hub, Guangzhou Baiyun Station. This involved the concentrated deployment of intelligent service robots covering five primary functions: passenger service, luggage transport, patrol service, cleaning, and aerial inspection.

14.4.1 Multi-Type Service Robot System

Guangzhou Metro piloted the application of the "Five Rams" series of intelligent robots at Guangzhou Baiyun Station, comprising:

(1) Yang Xiao Ying (Welcoming Lamb): A customer service robot that transcends the spatial limitations of traditional service counters. Equipped with a 19-inch high-definition display screen, it integrates voice guidance and real-time speech interaction. It can not only guide passengers through security checks and promote civilized boarding but also provide convenient services like voice inquiries and information lookup.

(2) Yang Xiao Yun (Transporting Lamb): A luggage robot specifically designed to address the challenge of passengers carrying large luggage items, fulfilling the need for large luggage and parcel transport services. Utilizing SLAM navigation (Simultaneous Localization and Mapping) and multi-sensor fusion technology, it can autonomously plan paths and intelligently avoid obstacles.

(3)Yang Xiao Xun (Patrolling Lamb):A patrol service robot equipped with a LiDAR navigation system, enabling all-weather dynamic inspection of equipment, facilities, and passenger flow. Fitted with dual "visible light + thermal imaging" cameras and a smoke detection system, it accurately identifies abnormalities such as water accumulation, smoke, or passenger falls, and relays this information in real-time to the station control room.

(4)Yang Xiao Jie (Cleaning Lamb):A cleaning robot integrating sweeping, mopping, and dust pushing functions. Innovatively equipped with an AI-powered intelligent inspection system, it can precisely identify types of ground stains and automatically match the appropriate cleaning solution. It maintains the cleanliness of the station concourse around the clock, providing consistent and high-standard environmental hygiene for station operations.

(5)Yang Xiao Fei (Flying Lamb):An aerial inspection robot employing "low-altitude + 3D" technology, carrying high-precision LiDAR and a thermal imaging gimbal. Using 3D modeling and panoramic aerial photography, it performs real-time video capture and dynamic flood monitoring at station entrances/exits, thereby enhancing the metro's integrated "ground-air" emergency response system layout.

These robots constitute a comprehensive station service robot system, covering the main operational scenarios of the station and establishing a multi-level service and monitoring network from the ground to the air.

14.4.2 Robot Collaborative Operation Mode and Effectiveness

The intelligent service robots of Guangzhou Metro do not operate in isolation; instead, they achieve collaborative linkage through the SuiTeng OS platform, forming an integrated solution. For instance:

When Yang Xiao Xun detects leftover trash on the ground during its patrol, it can notify Yang Xiao Jie to proceed to clean the area.

When Yang Xiao Ying identifies a passenger requiring luggage 搬运 assistance, it can summon Yang Xiao Yun to provide help.

When Yang Xiao Fei monitors passenger flow congestion at an entrance/exit, it can alert Yang Xiao Ying to go and assist with crowd guidance.

This inter-robot collaboration significantly expands the capability boundaries of individual robots, creating a synergistic effect where the whole is greater than the sum of its parts (1+1>2). By sharing location information, task status, and environmental data, the robot cluster achieves efficient task allocation and coordinated execution.

Guangzhou Metro has indicated that based on the pilot results of the robot matrix at Guangzhou Baiyun Station, it will continue to debug and upgrade the robots according to the specific demands of the metro environment. This involves constantly refining service details and broadening the scope and depth of applications. The aim is to provide passengers with a smarter and more convenient travel experience and deliver more efficient management for station operations, ultimately establishing Guangzhou Baiyun Station as a "Panoramic Smart Service Station." This effort promotes the deep application and integrated innovation of intelligent technology within metro operation and service scenarios.

The application of intelligent service robots not only elevates station service quality and operational efficiency but also reduces repetitive and labor-intensive manual tasks. This allows station staff to focus on more complex and value-added work. Furthermore, the innovative use of these robots enhances Guangzhou Metro's technological image and offers passengers a novel travel experience.

14.5 International Vocational Education System for Intelligent Stations and Smart Operation & Maintenance

To facilitate the international promotion of Guangzhou Metro's Intelligent Station and Smart O&M technologies, it is essential to establish a systematic vocational education and training system. This system will cultivate international technical talents equipped to meet the development requirements of intelligent metro systems.

14.5.1 Vocational Education Framework and Content Design

The international vocational education system targeting Intelligent Station and Smart O&M technologies should be competency-based, with training content designed across four progressive levels: "Fundamental Theory – Technical Skills – Practical Application – International Perspective":

(1) Fundamental Theory Module: Covers core theoretical knowledge including Intelligent Station system architecture, Smart O&M principles, Internet of Things (IoT) technology, and Big Data analytics, laying the foundation for subsequent technical learning.

(2) Technical Skills Module: Focuses on specific technologies for Intelligent Stations and Smart O&M, offering specialized tracks such as SuiTeng OS platform operation and development, intelligent equipment installation and debugging, data acquisition and analysis, and fault diagnosis and handling.

(3) Practical Application Module: Develops trainees' ability to solve practical problems through real-world case studies and hands-on training. This includes designing intelligent passenger flow control schemes, formulating equipment predictive maintenance strategies, and developing intelligent emergency response

plans.

(4) International Perspective Module: Introduces international standards, global technology trends, and typical international case studies. This cultivates a global mindset and cross-cultural communication skills among trainees, promoting the international application of Guangzhou Metro's technologies.

This training system should be implemented through diverse formats including classroom teaching, virtual simulation, practical hands-on training, and field internships, ensuring trainees can effectively translate theoretical knowledge into practical working capabilities.

14.5.2 Specialized Skills Training for Smart O&M

Reflecting the characteristics of Smart O&M technologies, vocational education should strengthen training in the following specialized skills:

(1) Data Acquisition and Processing: Learn the principles, installation, and debugging methods of various sensors; master fundamental skills in data acquisition, cleansing, and processing.

(2) Equipment Condition Monitoring and Assessment: Master methods for assessing equipment health status based on monitoring data; learn equipment fault prediction techniques and remaining useful life prediction models.

(3) Predictive Maintenance Strategies: Understand methods for developing maintenance plans based on equipment condition predictions; learn techniques for optimizing the allocation of maintenance resources.

(4) Smart O&M Platform Operation: Master the operational procedures of the Smart O&M platform; learn processes for work order management, maintenance process monitoring, and maintenance quality evaluation.

(5) Cloud-Edge Collaborative System O&M: Learn the principles and operational maintenance methods of the cloud-edge collaborative architecture; master configuration and management techniques for edge devices.

This specialized skills training should be closely integrated with Guangzhou Metro's practical cases, such as the intelligent maintenance system for Line 10 trains and the cloud-edge collaborative signaling intelligent maintenance system. This enables trainees to grasp the core skills of Smart O&M by studying real-world applications.

14.5.3 International Curriculum and Teaching Resource Development

To advance the international application of Guangzhou Metro's Intelligent Station and Smart O&M technologies, a series of international curricula and teaching resources need development:

(1) Bilingual Curriculum System: Develop a bilingual training curriculum system based on the technical practices of Guangzhou Metro's Intelligent Stations and Smart O&M, including course syllabi, textbooks, and lab guides.

(2) Virtual Simulation Platform: Develop a virtual simulation platform for Intelligent Stations and Smart O&M, enabling virtual training for practical skills like system configuration, equipment debugging, and fault handling.

(3) Training Base Construction: Construct a comprehensive Intelligent Station and Smart O&M training base replicating Guangzhou Metro's real environment, providing an authentic setting for practical operations.

(4) Teaching Faculty Development: Cultivate a team of "dual-qualified" instructors possessing both theoretical teaching ability and field experience, including technical experts from Guangzhou Metro.

(5) International Certification System: Collaborate with internationally recognized certification bodies to develop an international certification system for Intelligent Station and Smart O&M technologies, enhancing the global recognition of the training.

The development of these teaching resources will provide robust support for the international promotion of Guangzhou Metro's technologies and cultivate technical talents suited for the international market.

14.5.4 Training Models and Assessment Certification

To ensure the quality of vocational education, innovative training models and assessment certification mechanisms are required:

(1) Blended Learning Model: Combine online learning with offline practice, utilizing technologies like virtual simulation and augmented reality to improve

training flexibility and effectiveness.

(2) Project-Driven Teaching: Use real projects as vehicles, organizing trainee teams to complete tasks such as Intelligent Station design and Smart O&M solution formulation, fostering comprehensive capabilities.

(3) Skills Certification System: Establish a tiered skills certification system, ranging from basic operations to system design, and from single skills to comprehensive application, providing a clear career progression path for trainees.

(4) International Joint Training: Collaborate with foreign metro companies and vocational education institutions to conduct joint training programs, promoting technical exchange and cultural integration.

(5) Continuous Improvement Mechanism: Implement training quality evaluation and feedback mechanisms to continuously optimize training content and methods, maintaining the relevance and practicality of the training.

Through these innovations in training models and assessment certification, the quality of vocational education can be assured, providing a talent guarantee for the international application of Guangzhou Metro's Intelligent Station and Smart O&M technologies.

Concluding Remarks

The development of Guangzhou Metro's Intelligent Stations and Smart O&M system, characterized by the continuous iteration of the SuiTeng OS platform, relentless innovation in the Smart O&M system, and the large-scale application of service robots, has achieved a comprehensive intelligent transformation of station operations and equipment maintenance. These technological innovations and practical experiences not only provide strong support for the high-quality development of Guangzhou Metro itself but also offer valuable experience for the intelligent upgrade of global urban rail transit. By constructing a systematic international vocational education system, these technological innovations and practical experiences can be applied and promoted more widely, thereby driving the intelligent development of urban rail transit worldwide.

Chapter 15: Emergency Management and Safety Assurance System

15.1 Emergency Management System Architecture and Operational Mechanism

Guangzhou Metro's emergency management system adopts a comprehensive management model characterized by "government leadership, corporate responsibility, departmental coordination, and societal participation." It has established a systematic, standardized, and regulated emergency management mechanism. Guided by the core document, the "Guangzhou City Urban Rail Transit Operational Emergency Contingency Plan," and adhering to the working principles of "people-oriented, science-based decision-making; unified command, tiered responsibility; rapid response, coordinated; combining normal and emergency operations, ensuring effective response," the system guarantees the comprehensiveness and effectiveness of emergency management efforts.

The Guangzhou Municipal Government has established the Emergency Command Headquarters for Handling Urban Rail Transit Operational Emergencies (hereinafter referred to as the Municipal Headquarters). The Chief Commander is the Vice Mayor in charge of transportation. Deputy Commanders include the Deputy Secretary-General of the Municipal Government assisting with transportation, the primary responsible persons from the Municipal Transportation Bureau, the Municipal Emergency Management Bureau, Guangzhou Metro Group, and the responsible person from the Municipal Public Security Bureau in charge of related affairs. The office of the Municipal Headquarters is located within the Municipal Transportation Bureau. It is responsible for daily coordination and management tasks, including proposing the initiation and termination of emergency responses, organizing the implementation of decision-making deployments, and aggregating and reporting incident information. This centralized and unified command structure ensures the efficiency and coordination of emergency responses.

Guangzhou Metro's emergency management system covers the entire process, from risk identification to emergency response. According to the "Guangzhou City Urban Rail Transit Operational Emergency Contingency Plan," operational risks are categorized into five major types:

• **Facility Monitoring and Maintenance Risks:** Risks related to structural deformation, settlement, or collapse of bridges, tunnels, tracks, roadbeds, etc.

• **Equipment Operation and Maintenance Risks:** Risks arising from equipment failures (in power supply, communications, signaling, rolling stock, electromechanical systems, etc.) leading to service disruptions, fires, etc.

• **Train Operation Organization Risks:** Risks resulting from non-standard dispatch command or train operation organization, potentially causing train derailments, conflicts, etc.

•**Passenger Service Organization Risks:** Risks of crowding and stampedes due to large passenger flows.

• **Operational Environment Risks:** Operational risks caused by external factors such as natural disasters and public security incidents.

This comprehensive risk categorization system provides a scientific foundation for precise prevention and control.

Table 15-1 Organizational Structure and Responsibility Assignment for Emergency Management in Guangzhou Metro

Organizational Unit	Primary Responsibilities	Constituent Departments / Composition
Municipal Command Headquarters	Overall command, coordination, and strategic decision-making for major emergencies.	Led by the Municipal Government, with core participation from the Transportation Bureau, Emergency Management Bureau, Public Security Bureau, and Guangzhou Metro

		Group, among other relevant municipal authorities.
On-Site Command Post	Directs and executes all on-site emergency response and rescue operations.	Typically integrates various functional teams: Comprehensive Coordination, Emergency Repair & Rescue, Firefighting & Rescue, Evacuation & Transport, Medical Rescue, and Security & Stability Maintenance.
Expert Support Group	Provides specialized technical analysis, decision-making consultation, and post-incident assessment.	Comprises senior technical experts from diverse fields relevant to rail transit emergencies (e.g., structural engineering, power supply, signaling, rolling stock).
Operating Entity (Guangzhou Metro Group)	Responsible for initial incident response, immediate protective actions, daily risk management, preventive maintenance, and implementation of directives from command headquarters.	Executed by Guangzhou Metro Group and its subsidiary operational units directly managing station and line operations.

Guangzhou Metro's emergency management adheres to the principle of "integrating peacetime preparedness with emergency response," closely linking daily prevention with emergency handling.

During the routine management phase, the focus lies on proactive risk prevention and control. This includes the systematic identification and mitigation of potential safety hazards, strengthened maintenance of equipment and facilities, and the regular conduct of emergency drills and training programs. These initiatives collectively enhance staff preparedness and emergency response capabilities.

When an incident occurs, the emergency response phase is promptly activated according to the established tiered response system. This triggers the rapid mobilization of necessary resources and the immediate implementation of on-site containment and rescue operations, aiming to minimize impacts and losses to the greatest extent possible.

In the post-incident phase, comprehensive efforts are undertaken. These include managing aftermath issues, conducting thorough incident investigations to determine the root cause, and performing detailed summary evaluations. The lessons learned are systematically integrated back into the emergency management framework, driving its continuous refinement and improvement.

15.2 Emergency Response Plans and Command System

Guangzhou Metro has established a comprehensive three-tier emergency response plan system, encompassing overall plans, specialized plans, and on-site disposal schemes. This system achieves complete coverage, extending "horizontally across all areas and vertically down to the grassroots level." Guided by the overarching Guangzhou City Urban Rail Transit Operational Emergency Contingency Plan, district-level plans (e.g., the Yuexiu District Urban Rail Transit Operational Emergency Contingency Plan) are formulated based on the municipal plan. Subsequently, individual stations and operating units develop specific on-site disposal schemes according to their actual conditions, forming a complete network of emergency management plans.

15.2.1 Modern Emergency Command System

Guangzhou Metro's emergency command system adopts a three-tiered structure: "Municipal Headquarters – On-site Command Post – Operating Unit." This framework leverages informatization and intelligent technologies to achieve rapid response and precise command.

The Municipal Headquarters is responsible for overall command and coordination.

The On-site Command Post is established based on the incident type and integrates specialized modules such as Comprehensive Coordination, Emergency Repair & Rescue, Firefighting & Rescue, Evacuation & Transport, Medical Rescue, and Security & Stability Maintenance. This enables a modular approach to emergency response.

The Operating Unit (Guangzhou Metro Group) is responsible for initial incident response and information reporting.

A key component of this system is the "Call-and-Respond" emergency command application, developed by the Municipal Public Security Bureau's Public Transport Sub-bureau. This application facilitates "one-click" mobilization of the station's minimum emergency unit, which includes police officers, auxiliary police, station staff, and security personnel. It effectively breaks down communication barriers between the police and the metro operator, enabling command issuance anytime, anywhere, simultaneous reporting of information, and requests for support, thereby providing crucial technological support for emergency handling. During an incident, the system instantly connects the affected metro station with the command center, the responsible police station, nearby support-capable police stations, and various specialized police units. Information such as the incident location and disposal instructions is delivered accurately within seconds, significantly improving response speed.

15.2.2 Emergency Command and Coordination Mechanism

Guangzhou Metro's emergency command system has established a "concentric circle" response model. This model centers on the affected station and involves the rapid assembly of police forces from adjacent police districts and stations, along with specialized units, forming a protective "golden bell cover" for the rail transit system.

Concurrently, the Public Transport Sub-bureau has conducted research on emergencies occurring within the Guangzhou Metro system over the past decade, categorizing them into ten major types, including incidents related to operational safety and natural disasters. Through targeted training drills and organized emergency response tests, they strive to achieve "one guideline per category," "one plan per station," and "one drill per month." This systematic approach promotes the professionalization, standardization, and enhanced efficiency of emergency response and incident handling operations.

15.2.3 Multi-Level Response Mechanism

Guangzhou Metro classifies emergency responses into four distinct levels based on the severity and impact of operational emergencies: Level I (particularly significant), Level II (major), Level III (relatively large), and Level IV (general). The criteria for this classification encompass multiple dimensions, including casualties, direct economic losses, and the duration of service disruption. For instance, a general operational emergency (Level IV) is defined as an incident resulting in fewer than 3 fatalities, or fewer than 10 severe injuries, or direct economic losses between RMB 500,000 and RMB 10 million, or a continuous service interruption lasting more than 2 hours but less than 6 hours.

Different response levels activate corresponding command hierarchies and determine the scope of the emergency response:

Level I and Level II responses are commanded and coordinated by the Municipal Command Headquarters.

Level III responses are coordinated by the Office of the Municipal Command Headquarters, which may escalate coordination to the full Municipal Command Headquarters if necessary.

Level IV responses are managed by the Operating Unit (Guangzhou Metro Group), which can request coordination support from the Office of the Municipal Command Headquarters if needed.

This tiered response mechanism ensures high-level coordination and resource allocation for major incidents while enabling rapid and efficient of routine, smaller-scale events at the operational level. This structured approach optimizes the use of emergency management resources across the spectrum of potential incidents.

15.3 Safety Assurance Technology System

Guangzhou Metro's safety assurance technology system is founded on the philosophy of "Prevention First, Combining Prevention and Response." It integrates various advanced technologies, including intelligent security screening, flood and typhoon prevention, and equipment monitoring, to construct a comprehensive, multi-layered safety defense line.

15.3.1 Intelligent Security Screening Technology

Guangzhou Metro's security screening process follows the procedure of "Guidance – Inspection – Identification – Disposal." It utilizes security equipment such as channel-type X-ray machines, walk-through metal detectors, handheld metal detectors, liquid scanners, and explosive trace detectors. Screening is conducted in accordance with the principle of "comprehensive screening, simultaneous inspection of persons and belongings, and strict inspection upon suspicion" for all individuals entering the station and their carried luggage. The Guangzhou City Urban Rail Transit Security Screening Work Regulations specify operational standards, such as conducting focused re-inspection of alarm-indicating areas on passengers who trigger the walk-through metal detector, thereby enhancing the targeting and effectiveness of security checks.

Guangzhou Metro emphasizes humanized management within its security operations. For special groups, including pregnant women, students under 1.3 meters tall wearing school uniforms and holding student discounted metro tickets, and persons with mobility challenges, alternative methods combining handheld metal detector scans and visual inspection may be applied as the situation warrants. This differentiated screening approach ensures security while demonstrating humanistic care. Furthermore, the regulations stipulate that if large passenger crowds congest the security screening area due to peak passenger flow, severe weather, or equipment failure, the operating unit must take effective measures to maintain order, strengthen passenger flow guidance, and, depending on the situation, implement measures such as passenger flow restrictions or station closures to ensure passenger safety.

15.3.2 Flood and Typhoon Prevention Technology

In response to the rainy and typhoon-prone characteristics of the Guangzhou region, Guangzhou Metro has established a comprehensive flood and typhoon prevention technology system. This system achieves full intelligent monitoring coverage, with automatic water level monitoring and alarm devices installed at the entrances/exits of 302 operational underground stations and 331 key tunnel sections. Additionally, rain gauges and video monitoring equipment are installed at entrances/exits to monitor external water accumulation in real-time. A "Three-Layer

Defense" system is established at entrances/exits, where secondary and primary warning lines, as well as a station closure line, are scientifically marked. Emergency measures are activated progressively based on the depth of water accumulation.

Guangzhou Metro has implemented a "5-Level Metro Line Suspension Mechanism," which includes: closing entrances/exits, closing the station, suspending a section, suspending a line, and suspending the entire network. According to this suspension mechanism, for elevated and surface sections, corresponding speed restrictions or operational suspensions are implemented when wind speed monitors detect different thresholds of maximum wind force. Should any of the above situations occur, Guangzhou Metro will inform passengers of service adjustments through metro TVs, station broadcasts, notices, official social media accounts, traffic radio, and media outlets, facilitating travel plan adjustments.

Furthermore, Guangzhou Metro has strengthened its physical flood prevention infrastructure by installing 2,113 flood barrier plates and constructing reinforced concrete retaining walls at 13 depot access lines, significantly enhancing physical flood resistance capabilities. Special contingency plans have been developed and revised, and 340 station flood response maps, 32 depot flood response maps, and 44 substation flood response maps have been created, realizing the concept of "One Station, One Strategy" and "One Depot, One Map." The metro is also equipped with advanced drainage equipment, including high-power pumps, hydraulic power stations, "Dragon Suction" pumps, and diesel self-priming pumps, with a total pumping capacity reaching 10,000 cubic meters per hour.

15.3.3 Equipment and Facility Safety Monitoring Technology

Guangzhou Metro has strengthened the monitoring and maintenance of critical equipment and facilities to ensure operational safety. For special equipment such as escalators and elevators, Guangzhou Metro strictly adheres to national regulations and engages qualified professional contractors for maintenance. In response to growing passenger volumes, Guangzhou Metro has intensified equipment inspection and patrol efforts to ensure all escalators and elevators remain in optimal working condition.

A structured maintenance regime is implemented for escalators, including:

- Daily inspections
- Bi-weekly routine maintenance
- Quarterly equipment adjustment and testing
- Semi-annual comprehensive maintenance
- Annual thorough overhaul

Regarding power supply system safety, Guangzhou Metro has deployed intelligent monitoring devices that track key electrical parameters in real-time. Through big data analytics, the system predicts potential equipment failure risks, enabling proactive maintenance.

To further enhance safety assurance, Guangzhou Metro has established:

- 29 specialized flood and wind emergency response teams totaling 559 members
- A 200-person network-wide emergency rescue corps

These teams undergo continuous professional development to maintain expertise in disaster response, providing robust organizational support for equipment and facility security. This comprehensive approach combines technical monitoring with human expertise to create a multi-layered safety protection system.

Table 15-2 Flood and Typhoon Prevention Technical Measures at Guangzhou Metro

Technical Category	**Specific Measures**	**Application Scope**	**Implementation Outcomes**
Intelligent Monitoring	Automatic water level monitoring and alarm devices; rain gauges; video surveillance equipment.	Entrances/exits of 302 operational underground stations and 331 key tunnel sections.	Enables real-time grasp of external water accumulation levels for timely decision-making.
Phased Service Suspension	5-level metro line suspension mechanism.	Applicable across the entire metro network.	Facilitates differentiated operational measures based on dynamically

			assessed risk levels.
Physical Protection	Installation of 2,113 flood barrier plates; construction of 13 reinforced concrete retaining walls.	Station entrances/exits and depot access lines.	Significantly enhances physical flood resistance and structural defense capabilities.
Emergency Drainage	Deployment of high-power pumps, "Dragon Suction" high-capacity pumps, and diesel self-priming pumps.	Flood-prone stations and tunnel sections.	Achieves a substantial total pumping capacity of 10,000 m^3/hour for rapid water discharge.
Precision Management	Implementation of "One Station, One Strategy" and "One Depot, One Map" protocols.	Development of 340 station-specific, 32 depot-specific, and 44 substation-specific flood response charts.	Ensures highly targeted and customized flood prevention measures for each critical asset.

15.4 Emergency Response and Handling Mechanism

Guangzhou Metro's emergency response and handling mechanism comprises interconnected phases including monitoring and early warning, information reporting, initial response, emergency response activation, on-site 处置, and post-incident management, forming a complete emergency management closed-loop system.

15.4.1 Emergency Response Process

Guangzhou Metro has established comprehensive monitoring, early warning, and information reporting mechanisms. Regarding monitoring and early warning, the operating unit and other relevant departments conduct preventive and monitoring activities, issuing timely early warning information. This information includes the category of the emergency, warning level, start time, potential impact area, key advisories, recommended actions, and the issuing authority. For information reporting, adhering to the principle of "Initial Report for Speed, Follow-up Reports for Accuracy, Final Report for Completeness," the operating unit must immediately report any discovered operational emergency to the Municipal Headquarters Office. The report content covers the incident's time, location, information source, nature, severity, impact scope, evolving trend, and measures already taken.

Upon activation of the emergency response, corresponding measures are implemented based on the incident level. For general operational emergencies (Level IV), the operating unit initiates and manages the response. For relatively large or more severe incidents, the Municipal Headquarters activates Level III, II, or I emergency responses accordingly. The 2023 case of a high-altitude fall accident at the Chisha Depot project of Guangzhou Metro Line 11 underscores the critical importance of standardized information reporting and emergency response. In this incident, the labor subcontractor failed to report as required, resulting in underreporting and subsequent serious disciplinary action.

15.4.2 On-site Disposal Measures

Guangzhou Metro implements specialized on-site disposal measures tailored to different types of operational emergencies. During the July 2021 water inflow incident at Shenzhou Road Station on Line 21, Guangzhou Metro activated its emergency response, ensured the timely and safe evacuation of passengers within the station, and leveraged its ground-underground coordinated response mechanism to involve traffic police and local police in rescue and crowd management efforts. This effective response resulted in no casualties. Normal operations were gradually restored after 8 hours of intensive rescue and handling efforts.

Key on-site disposal measures primarily include:

• **On-site Evacuation:** Formulating evacuation routes promptly based on the emergency situation and passenger needs, establishing temporary evacuation areas or emergency shelters.

• **Personnel Search and Rescue:** Utilizing technologies like personnel location and mobile reconnaissance to pinpoint trapped individuals, and deploying specialized teams such as Emergency Repair & Rescue and Firefighting & Rescue groups for search and rescue operations.

• **Passenger Transfer:** Organizing and deploying transportation resources like buses, establishing emergency bus bridging services to evacuate large passenger flows.

- **Traffic Management:** Maintaining traffic order at the incident site and surrounding areas, setting up traffic cordons, and implementing traffic controls to prevent secondary incidents.

- **Medical Rescue:** Dispatching medical rescue teams to the scene, setting up temporary medical points, and providing treatment for the injured.

- **Emergency Repair and Rescue:** Mobilizing engineering rescue teams to conduct urgent repair and rescue operations on affected facilities and equipment.

15.4.3 Post-Incident Management and Recovery Mechanism

Following the conclusion of emergency response operations, Guangzhou Metro implements a systematic post-incident management process. This encompasses:

- **Aftermath Handling:** Properly resolving conflicts and disputes arising from the incident response; organizing the restoration of commandeered materials and facilities; managing medical assistance and compensation for the injured.

- **Incident Investigation:** Establishing an investigation team according to the incident level and relevant regulations to determine the sequence of events, causes, casualties, and direct economic losses. The team ascertains the nature of the incident and responsibility, proposing handling recommendations and preventive/rectification measures.

- **Response Evaluation:** After the emergency response concludes, the Municipal Headquarters organizes an assessment of the emergency handling process, summarizing lessons learned and proposing improvement measures.

In the handling of the high-altitude fall accident at the Chisha Depot of Guangzhou Metro Line 11, the investigation team determined it was a general production safety liability accident. The primary causes were workers failing to wear and use safety harnesses as required, and personnel operating special equipment without the requisite qualifications. The investigation team recommended lawful penalties against the responsible unit, Shenzhen Chenxi Company, and its legal representative. It also advised that the Municipal Transportation Bureau handle the regulatory violations discovered during the investigation involving the metro consulting company and China Railway Construction Engineering Group according to regulations. This rigorous post-incident management mechanism provides a basis for accountability and systematic improvement.

15.5 International Vocational Education System for Emergency Management and Safety Assurance

To cultivate international technical talents meeting the requirements of Guangzhou Metro's emergency management and safety assurance, it is essential to construct a systematic vocational education and training system. This system will transform Guangzhou Metro's advanced experience and technologies into transferable educational resources.

15.5.1 Training Content System

The international vocational education system for emergency management and safety assurance should be competency-based, with training content designed across four progressive levels:

(1) Fundamental Theory Module: Covers basic theoretical knowledge including emergency management principles, safety assurance systems, risk assessment methodologies, and relevant laws and regulations, laying the foundation for subsequent technical learning.

(2) Technical Skills Module: Focuses on specific technologies for emergency management and safety assurance, offering specialized tracks such as emergency plan development, emergency command system operation, intelligent security screening technology, flood and typhoon prevention technology, and equipment monitoring technology.

(3) Practical Application Module: Develops trainees' ability to solve practical problems through real-world case studies and hands-on training, including designing emergency drills, formulating emergency response plans, and optimizing safety assurance measures.

(4) International Perspective Module: Introduces international standards, global

technology trends, and typical international case studies, cultivating a global mindset and cross-cultural communication skills to promote the international application of Guangzhou Metro's emergency management and safety assurance technologies.

This training system should be implemented through diverse formats including classroom teaching, virtual simulation, practical hands-on training, and field internships, ensuring trainees can effectively translate theoretical knowledge into practical working capabilities.

15.5.2 Training Methodology System

Vocational education for emergency management and safety assurance at Guangzhou Metro should employ diversified training methods, including:

(1) Case-Based Teaching: Uses real Guangzhou Metro cases (e.g., the 2021 Shenzhou Road Station water inflow incident, the 2023 Chisha Depot fall accident) as teaching materials to cultivate trainees' problem analysis and resolution skills.

(2) Simulation Drills: Utilizes virtual simulation technology to create virtual environments for metro emergency command and disposal, allowing trainees to make decisions and conduct command within simulated scenarios, enhancing their emergency response capabilities.

(3) Live Exercise Training: Conducts emergency drills within the actual Guangzhou Metro environment, such as the "Monthly Drill" activities organized by the Public Transport Sub-bureau. This provides trainees with firsthand experience of the emergency response process, fostering the development of "muscle memory."

(4) Collaborative Learning: Employs group discussions and team projects to cultivate trainees' teamwork, communication, and coordination skills, preparing them for the multi-departmental coordination demands of emergency management.

The "Elite Troops & Strong Stations – Minimum Combat Unit" large-scale training, competition, and improvement campaign by the Public Transport Sub-bureau is a prime example of live exercise training. By enhancing police-enterprise and ground-underground coordination, it refines emergency response tactics, issues disposal guidelines clarifying processes and standardizing legal language, and utilizes methods like "Red vs. Blue" exercises to elevate readiness and instill "muscle memory" throughout the teams.

15.5.3 International Certification System

To promote the international application of Guangzhou Metro's emergency management and safety assurance technologies, an international certification system needs establishment:

(1) Skills Standard Development: Develop internationally recognized skills standards based on Guangzhou Metro's technical practices in emergency management and safety assurance, including competency standards, training standards, and assessment criteria.

(2) Certification Body Collaboration: Partner with internationally renowned certification bodies to develop an international certification system for emergency management and safety assurance technologies, enhancing the global recognition of the training.

(3) Bilingual Teaching Resources: Compile English-language training materials, produce bilingual instructional videos, and develop multilingual virtual simulation platforms to meet international training needs.

(4) International Exchange Mechanisms: Collaborate with foreign metro companies and vocational education institutions on initiatives such as faculty exchanges, student swaps, and joint training programs to promote technical exchange and cultural integration.

Through this comprehensive vocational education and training system, Guangzhou Metro cultivates a large number of technical and skilled talents in emergency management and safety assurance. This not only meets its own development needs but also provides a reference for talent cultivation in other city metro systems, thereby promoting the international application of Guangzhou Metro's technologies in this field.

Concluding Remarks:

Guangzhou Metro's emergency management and safety assurance system, through its robust organizational structure, systematic contingency plans, advanced technical equipment, and professional vocational education system, achieves full-process management from risk prevention to emergency response. These experiences hold significant reference value for cities worldwide that are constructing or expanding their urban rail transit systems, and also provide rich teaching resources and practical cases for international vocational education in the rail transit sector. As technology continuously evolves and emergency management concepts keep innovating, Guangzhou Metro will persist in refining its emergency management and safety assurance system, offering Chinese solutions for the safe operation of global urban rail transit.

Part IV: Intelligent Innovation Technologies

Chapter 16: Overview of Intelligent Metro Development

16.1 Development Background and Strategic Significance of the Intelligent Metro

Guangzhou Metro's intelligent transformation stems from a triple driver of urban development demands, policy support, and technological innovation. As a core city within the Guangdong-Hong Kong-Macao Greater Bay Area (GBA), Guangzhou's urban rail transit system faces unprecedented passenger flow pressure and operational complexity. According to statistical data, Guangzhou Metro's average daily passenger volume grew from an initial 172,000 passengers to 9.06 million passengers per day in 2019, representing an increase of over 50 times. In 2020, Guangzhou Metro's passenger intensity reached 12,400 passengers per day per kilometer, ranking highest nationally, with the peak single-day passenger flow even reaching 11.13 million. This high-intensity passenger load places extreme demands on metro operational efficiency, safety assurance, and service quality.

From the policy perspective, the national strategic level provides clear direction for the development of the intelligent metro. The Outline Development Plan for the Guangdong-Hong Kong-Macao Greater Bay Area explicitly proposes the general requirements to "build a fast transportation network for the Greater Bay Area" and "accelerate the construction of intercity railways". The Guangdong Province *14th Five-Year Plan for the Development of the Comprehensive Transportation System* further sets specific goals, aiming to basically form the layout of a comprehensive three-dimensional transportation network by 2025. The white paper Innovation and Development of Urban Rail Transit in the New Era, released by Guangzhou Metro Group, outlines the enterprise-level blueprint for "Intelligent Rail Transit": user-centric, powered by a new generation of artificial intelligence technologies, and building a data-driven advanced urban rail transit information integration system architecture.

Technological innovation serves as the third major driver for intelligent metro development. The maturation of emerging technologies such as cloud computing, the Internet of Things (IoT), Artificial Intelligence (AI), and 5G has made the intelligentization of rail transit possible. Guangzhou Metro recognized the importance of technological innovation early on, undertaking a key National 863 Program project in 2011 to conduct research on key technologies for urban rail train in-transit monitoring and safety early warning. In March 2016, the National Development and Reform Commission approved the establishment of the "National Engineering Laboratory for System Safety and Operation Maintenance Assurance of Urban Rail Transit", led by Guangzhou Metro Group. This laboratory has become a crucial technical support pillar for Guangzhou Metro's intelligent transformation.

The development of the intelligent metro holds profound strategic significance for Guangzhou Metro. Firstly, intelligentization is an inevitable choice for managing an ultra-large-scale network. As the scale of Guangzhou Metro's network continuously expands – growing from 18.5 kilometers in 1997 to 14 lines totaling 531 kilometers (data as of 2021), an increase of approximately 28 times – traditional manual management methods can no longer adequately meet operational demands. Secondly, the intelligent metro is a key pathway to enhancing the passenger travel experience. Through intelligent technologies, it enables precise, convenient, and all-time, all-space passenger services, allowing the metro to proactively engage with passengers on a deeper level. Thirdly, the intelligent metro represents a strategic high ground for leading industry development. Through its intelligent transformation, Guangzhou Metro not only elevates its own operational standards but also exports technologies, standards, and integrated solutions domestically and internationally, achieving a leap from "catching up" to "running alongside" and then to "leading the pack".

Table 16-1 Multidimensional Driving Factors for the Development of Guangzhou Metro's Intelligent Metro System

Driving Dimension	Specific Manifestations	Impact Areas
Urban Development Demands	Extremely high passenger volume with average daily flow exceeding 10 million; highest passenger intensity nationwide.	Places critical demands on operational efficiency, safety assurance, and service quality, necessitating intelligent solutions for management and service delivery.
Policy Support and Guidance	Top-level directives including the Guangdong-Hong Kong-Macao Greater Bay Area Development Plan, the national Transport Power Strategy, and the *14th Five-Year Plan*.	Drives infrastructure construction, guides and funds technology research & development, and influences the formulation of industry standards for intelligent rail transit.
Technological Innovation Push	Maturation and convergence of key enabling technologies such as 5G, Artificial Intelligence (AI), Internet of Things (IoT), and Digital Twin.	Enables breakthroughs and widespread application in core areas including intelligent maintenance, smart passenger services, and advanced dispatch and control systems.
Corporate Development Objectives	Strategic goals to build a world-class rail transit brand and successfully implement the "Going Global" strategy.	Motivates internal management model innovation, facilitates the export of technologies and

	standards, and promotes the construction of a collaborative industrial ecosystem.

16.2 Development History and Phase Division of the Intelligent Metro

The intelligentization process of Guangzhou Metro was not achieved overnight; rather, it underwent a gradual evolution from digitalization to intelligentization, and further to integrated smart application. Based on its technological development characteristics and application depth, this process can be divided into three main stages.

Phase 1 (Prior to 2011): Digitalization Initiation and Foundation Building. This stage was characterized by infrastructure construction and the application of automated systems. From its early construction phase, Guangzhou Metro emphasized technological innovation. Progressing from importing advanced foreign technology for Line 1 to achieving a 70.21% localization rate for Line 2, the cost per kilometer decreased from RMB 660 million for Line 1 to RMB 480 million. During this phase, Guangzhou Metro achieved several technological breakthroughs: it was the first in China's industry to install Platform Screen Doors (PSDs); it established theories and methodologies for shield construction during line development; and it built foundational information systems and automated control systems, laying the groundwork for subsequent intelligent upgrades.

Phase 2 (2011-2018): Intelligentization Exploration and System Construction. A landmark event marking this stage was Guangzhou Metro undertaking a key National 863 Program project, conducting research on "Key Technologies for In-Transit Monitoring and Safety Early Warning of Urban Rail Trains." This project involved installing sensors on trains, developing wireless train-to-ground transmission equipment, establishing ground-based data transmission centers, and creating an intelligent system capable of monitoring train status in real-time. In 2016, Guangzhou Metro was approved to lead the establishment of the "National Engineering Laboratory for System Safety and Operation Maintenance Assurance of Urban Rail

Transit," constructing a "1+7+N" industry-university-research-application innovation ecosystem. During this period, Guangzhou Metro achieved breakthroughs in multiple technological fields: it independently developed and applied China's first linear motor train with complete independent intellectual property rights, the MTC-I type CBTC signaling system, and the Ultra-High-Speed Wireless Communication (EUHT) technology; it pioneered the "Hengdunni" auxiliary method for pressurized shield chamber entry nationally; it led the nationwide widespread application of diversified payment systems and broadband LTE wireless systems; and it was the first to achieve full life-cycle application of BIM technology in urban rail transit.

Phase 3 (2019-Present): Smart Integration and Ecosystem Formation. Marked by the release of Tencent OS 1.0 in 2019, Guangzhou Metro entered a phase of comprehensive smart development. SUNTING OS is the industry's first new-generation intelligent rail transit operating system based on the Industrial Internet and IoT. In 2021, Tencent OS was upgraded to version 2.0 and deployed on Guangzhou Metro Lines 18 and 22. At this stage, Guangzhou Metro's smart applications demonstrated characteristics of full business coverage, deep integration, and collaborative ecosystem building. In smart services, features like face recognition for fare payment, intelligent customer service, and onboard crowd density displays were introduced. In intelligent maintenance, a tunnel comprehensive inspection system was adopted, improving efficiency approximately 120-fold compared to manual inspection. In smart construction, intelligent metro requirements were incorporated into the design of new lines and the renovation of existing lines. In 2023, Guangzhou Metro further proposed the goal of building the "New Era Metro," striving to double the network mileage to 800 kilometers by 2023 and continuously advancing towards a more integrated, efficient, and passenger-centric intelligent rail transit system.

Figure 16-2 Guangzhou Metro's Intelligent Development Journey

Guangzhou Metro's Intelligent Development Journey		
Phase 1 (Pre-2011)	Phase 2 (2011-2018)	Phase 3 (2019-Present)

Digitalization Initiation	Intelligentization Exploration	Smart Integration & Ecosystem
•Construction of foundational automated control systems for core operations •Achieved breakthroughs in technology localization, significantly reducing costs •Introduced and assimilated key foreign technologies, notably Platform Screen Doors •Established initial information management and operational technology platforms	•Execution of the pivotal National 863 Program R&D project for train monitoring •Approval and establishment of the National Engineering Laboratory for safety and O&M •Attained multiple core technological breakthroughs (e.g., indigenous CBTC, EUHT) •Initiated R&D and deployment of integrated intelligent maintenance systems , indigenous CBTC, EUHT) •	•Successful launch and iterative upgrade of the Tencent OS platform (v1.0 to v2.0) •Achieved comprehensive coverage of intelligent applications across all operational scenarios •Established a collaborative ecosystem with industry partners for co-creation •Deepened the digital transformation of management, services, and maintenance

Guangzhou Metro's intelligent transformation embodies an evolutionary logic characterized by the progression from discrete technological breakthroughs to the construction of holistic systemic capabilities, transitioning from localized application to global optimization, and shifting from a technology-driven approach to a business value-oriented paradigm. This incremental and deliberate pathway not only aligns with the inherent patterns of technological maturation but also closely corresponds to the practical operational demands of metro systems. As such, it offers a valuable reference for the intelligent transformation of urban rail transit in other cities.

16.3 Core Technology System and Innovative Applications of the Intelligent Metro

Centered on the three core domains of "Smart Services, Intelligent Maintenance, and Smart Management," Guangzhou Metro has established a comprehensive technological system and realized numerous innovative applications, forming an industry-leading intelligent metro solution.

16.3.1 Tencent OS System Platform

SuiTang OS is the core intelligent metro platform jointly developed by Guangzhou Metro and Tencent. As the nation's first "Intelligent Brain" for rail transit, it fundamentally redefines the design philosophy of traditional industrial control systems. SuiTang OS has evolved from version 1.0 to 2.0: Version 1.0 was launched in 2019 at Canton Tower Station, establishing the platform's foundational architecture; Version 2.0 was released in 2021 and deployed for demonstration operations on Lines 18 and 22, significantly enhancing IoT capabilities and integrated data processing.

The technical architecture of Tencent OS comprises several core platforms: the IoT Platform enables unified device access and management; the Strategy Engine Platform supports intelligent linkage for diverse business scenarios; the Big Data Platform provides data storage, processing, and analytics capabilities; the Algorithm Platform integrates various AI algorithms; and the Open Platform attracts ecosystem partners to participate in application development. This architecture facilitates intelligent perception, data interaction, intelligent analysis, and coordinated operation of various station equipment systems, providing "coordinated and consistent" operational support for station services.

The application effectiveness of Tencent OS is remarkable. On Guangzhou Metro Line 18, the system enables intelligent passenger flow guidance. Through the interconnection of sensing devices, surveillance systems, fare gates, elevators, broadcasts, and mobile apps, it analyzes the distribution of crowd density on platforms in real-time and automatically guides passengers to less crowded waiting areas. For flood prevention safety, the system utilizes liquid level sensors installed at risk points to monitor station water infiltration in real-time. Once water levels reach a predefined warning threshold, the station's flood emergency plan can be activated with a single click, rapidly coordinating station equipment to guide passenger evacuation orderly. In station management, leveraging the Strategy Engine Platform of Tencent OS, scenario-based mode controls have been established, including one-click station opening/closing, video-based patrol inspection, and one-click activation of station emergency plans. This has enabled mobile station management and reduced operational costs.

16.3.2 Intelligent Maintenance System

Guangzhou Metro has built an advanced intelligent maintenance system, achieving a transition from "preventive maintenance" to "predictive maintenance." In tunnel inspection, the first tunnel comprehensive detection system developed by Guangzhou Metro can perform sub-millimeter level tunnel inspections at speeds of 60 km/h, improving efficiency approximately 120-fold compared to manual inspection. This system integrates technologies such as LiDAR scanning, 3D structured light measurement, and high-resolution line-scan camera imaging, enabling precise detection of defects like tunnel cracks, spalling, water leakage, and segment misalignment, with a maximum accuracy of 0.3 mm.

In rolling stock maintenance, the "Train 360-degree Inspection System" deployed on Guangzhou Metro Line 21 has gradually replaced about 80% of manual inspections. The "Jia Yi Wei" intelligent maintenance system, developed by PDT and piloted on the second phases of Lines 13 and 14, enables 7x24-hour "precise condition monitoring" of equipment and facilities. It facilitates 超前 fault prediction and accurate condition assessment, shifting the paradigm from reactive repair to precise "condition-based maintenance."

In communication technology, Guangzhou Metro leads the national key R&D project "New Generation Communication and Network Application Demonstration for Typical Industries," developing the "5G Train-borne Edge Gateway Equipment (5G TAU)" and the "Metro Intelligent Decision-making Platform." The 5G TAU is the world's first industry application terminal for the rail transit sector based on multi-network access and multi-service bearing, providing access capabilities for 5G, Wi-Fi, LTE-M, EUHT, and TETRA networks. The Metro Intelligent Decision-making Platform leverages technologies like comprehensive IoT sensing, data fusion analysis, and multi-dimensional digital twins to achieve convenient passenger services and panoramic control and management of stations and trains.

16.3.3 Smart Service Innovations

Guangzhou Metro has implemented numerous innovative applications in smart services, significantly enhancing the passenger travel experience. In ticketing services, face recognition for fare gate transit was introduced, piloted at Canton Tower Station on the APM Line and subsequently expanded on Lines 18 and 22. Passengers simply need to register with their real name and bind a payment method in the Guangzhou Metro App to pass through the gates seamlessly ("no sense" passage).

Regarding passenger information services, Line 18 features an embedded Platform Display System integrated directly into the Platform Screen Doors. It provides passengers with comprehensive information via video, including express/local train status, current station, next station, terminal station, onboard crowd density, and current waiting position. Inside the trains, intelligent circular lighting adjusts its brightness based on the level of passenger crowding. Utilizing vehicle load information, the system analyzes the passenger crowd density within each carriage and updates this information in real-time on the display screens at the next station's platform screen doors, assisting passengers in choosing less crowded carriages.

For customer service, all stations across the network are equipped with Smart Customer Service Centers supporting functions like ticket processing, ticket inquiry, facial registration, and electronic invoices. In 2025, at Yuejialu Station and Yunxiao Road Station on the second phase of Line 14, intelligent customer service robots based on large language models were officially deployed. These robots can guide passengers through voice and natural body movements and answer travel inquiries in real-time.

16.3.4 Smart Infrastructure

Guangzhou Metro has achieved significant innovations in the intelligentization of its infrastructure, enhancing safety, efficiency, and sustainability across its operations.

- **Escalator Systems:**

Escalators in intelligent demonstration stations are equipped with a smart operation and maintenance system. This system enables the real-time monitoring of operational status, provides immediate fault alerts, and offers health diagnostics alongside maintenance recommendations. Furthermore, escalators network-wide are capable of identifying passenger flow density and detecting specific passenger behaviors. They are also equipped with active protection features, such as emergency stop functionality, which contributes to improved operational safety.

• **Platform Screen Door (PSD) Systems:**

PDT Science & Technology has applied its self-developed, compatible Platform Screen Door Door Control Unit (DCU). This solution effectively addresses challenges related to replacing components from different PSD brands and the procurement difficulties of imported spare parts, thereby enhancing maintenance efficiency and system interoperability.

For obstacle detection in PSDs, intelligent stations utilize a system based on the Time of Flight (TOF) principle. It employs overhead, integrated transceiver LiDAR sensors to monitor the gap between the train doors and the platform screen doors. This advanced detection technology significantly reduces the probability and associated risks of entrapment incidents.

• Energy Management:

An intelligent Environmental Control and Energy Saving System, based on hybrid models and AI algorithms, has been implemented. This system utilizes Digital Twin and deep learning technologies to perform real-time optimization of the Environmental Control System (ECS) within metro stations. By dynamically adjusting system parameters based on real-time conditions and predictive analytics, it unlocks deep energy-saving potential, leading to substantially reduced operational energy consumption.

Table 16-3 Comparative Analysis of SuiTeng OS System Development Stages at Guangzhou Metro

System Characteristic	SuiTeng OS 1.0	SuiTeng OS 2.0	Future Evolution Direction
Release Year	2019	2021	Continuous Iteration

Application Lines / Deployment	Canton Tower Station, Tianhe Intelligent City Station	Lines 18 & 22	Network-wide Promotion & Deployment
Technical Features	Established platform foundation; achieved line-level application capabilities	Enhanced IoT capabilities; realized integrated data processing and analysis	Deep AI integration; autonomous decision-making capabilities
Core Functions	Comprehensive information display, intelligent passenger flow guidance, smart security, intelligent station management	Intelligent passenger flow guidance, intelligent station affairs, smart security, intelligent dispatch, automated vehicle maintenance	Panoramic intelligent operations; proactive service delivery
IoT Capability	Limited device connectivity and integration	Established unified IoT standards; achieved device interconnectivity and interoperability	Network-wide intelligent device perception and management
Ecosystem Model	Initial collaborative partnerships	Comprehensive open-sharing platform, actively engaging industry partners	Ecosystem co-prosperity; industrial collaborative innovation

16.4 System Architecture and Standardization Development of the Intelligent Metro

Throughout the development of its intelligent metro system, Guangzhou Metro has progressively established a comprehensive system architecture and a set of standards and specifications, providing a solid foundation for the sustainable advancement of intelligent metro initiatives.

16.4.1 Platform Architecture and Technical Characteristics

The intelligent metro system of Guangzhou Metro employs a layered architecture, comprising, from bottom to top, the Perception Layer, Network Layer, Platform Layer, and Application Layer.

- **Perception Layer: This** foundational layer deploys various sensors, intelligent cameras, smart terminals, and other devices to collect real-time information on the station environment, equipment status, and passenger flow data.

- **Network Layer: A** converged wired and wireless network architecture is utilized. The application of 5G technology, in particular, provides a more efficient solution for train-to-ground communication, supporting high-bandwidth, low-latency data transmission.

- **Platform Layer: The** five core platforms of SuiTeng OS 2.0 – the IoT Platform, Strategy Engine Platform, Big Data Platform, Algorithm Platform, and Open Platform – collectively form the "Digital Backbone" or core nervous system of the intelligent metro.

- **Application Layer: A** series of intelligent applications are realized at this top layer, including intelligent passenger flow guidance, intelligent station management, smart security, intelligent dispatch, and more.

The Guangzhou Metro intelligent platform exhibits three key technical characteristics:

Openness and Evolvability:Tencent OS is based on a componentized, microservices architecture, supporting the rapid development and iteration of business applications.

- **Intelligent Linkage: Through** the Strategy Engine Platform, coordinated control and emergency among multiple systems are achieved.

- **Data-Driven Operation:** Leveraging the Big Data Platform and Algorithm Platform, the system enables integrated data analysis and intelligent decision-making.

16.4.2 Standard System and Specification Development

Standardization is crucial for the scaled development of the intelligent metro. Guangzhou Metro actively participates in the formulation of industry standards, having undertaken the compilation of 28 industry standards over a recent five-year period. Within the "Intelligent Metro Function Grade" system released by Guangzhou

Metro, the intelligence level is classified into 4 grades, with Grade 1 being the basic level and Grade 4 the highest. This classification covers six key areas: passenger service, Train Operation Organization, Dispatch and Command, station management, Operation and Maintenance, and Safety Assurance & Emergency Response.

Regarding data standards, Tencent OS provides specifications for IoT access, big data, and service openness for urban rail transit. These standards facilitate the interconnection and interoperability of rail transit equipment and systems, effectively addressing the traditional problem of "Data Silos" between various subsystems. In terms of interface standards, Guangzhou Metro has promoted the unification of intelligent terminal device interfaces, providing technical specifications for the integration of equipment from different manufacturers.

An innovative approach in Guangzhou Metro's standardization efforts is using demonstration projects to drive standard formation. The demonstration applications at Canton Tower Station and Tianhe Intelligent City Station, followed by the network-wide promotion on Lines 18 and 22, provided real-world scenarios for validating and optimizing various intelligent metro standards. This pathway of "Practice-Summarization-Standardization" ensures the practicality of the standards while also accelerating the dissemination and adoption of the technologies.

16.5 Future Development Trends and Challenges of the Intelligent Metro

With continuous technological advancements and evolving operational demands, Guangzhou Metro's intelligent development faces new opportunities and challenges. A comprehensive analysis of technological evolution patterns and industry development trends indicates the following future directions for the intelligent metro.

16.5.1 Technological Development Trends

Artificial Intelligence (AI) technology will become more deeply integrated into all business domains of metro operations. Intelligent customer service systems based on Large Language Models (LLMs) have already been piloted on the second phase of Guangzhou Metro Line 14 and are expected to be extended across the entire network, enabling more natural and precise human-computer interaction. Digital Twin technology will progress from localized applications to full-network modeling, achieving real-time mapping and interactive optimization between the physical and digital metros. Self-learning algorithms will be gradually applied to equipment predictive maintenance, analyzing historical data and real-time status to identify potential equipment faults in advance, thereby shifting the paradigm from "periodic maintenance" to "condition-based maintenance."

Communication technology will undergo continuous upgrades. The comprehensive application of 5G technology will further enhance train-to-ground communication capabilities. Guangzhou Metro has already conducted research on various station and train-to-ground 5G applications on Line 18, including video surveillance, ticketing, intelligent escalator monitoring, and onboard intelligent maintenance. In the future, 5G networks will carry more critical services, supporting application scenarios such as ultra-high-definition video transmission, real-time monitoring, and remote control.

16.5.2 Operational Model Transformation

The construction of the intelligent metro will drive profound changes in operational models. In passenger service organization, the shift will be from "reactive response" to "proactive prediction." Guangzhou Metro is developing passenger flow prediction algorithms capable of forecasting short-term passenger flow changes, guiding entering passengers to better manage their time. In the future, leveraging big data and AI technologies, metro operators can more accurately predict passenger flow distribution, dynamically adjust transportation plans, and achieve precise matching of capacity and demand.

In maintenance models, the transition will be from "scheduled maintenance" to "predictive maintenance." The "Jia Yi Wei" intelligent maintenance system by PDT has already demonstrated the capability for 7x24-hour "precise condition monitoring" of equipment and facilities on the second phases of Lines 13 and 14. Looking ahead, with advancements in sensor technology and diagnostic algorithms, equipment health monitoring will become more accurate, and maintenance decisions will become more scientific, significantly improving equipment reliability and service life.

16.5.3 Regional Coordinated Development

The development of the Guangdong-Hong Kong-Macao Greater Bay Area (GBA) provides a broader context for Guangzhou Metro's intelligent development. Guangzhou Metro is preparing for or undertaking the operation and management of several intercity railways, such as the Guangqing, Xinbai-Guang, and Zhuhai-Zhuhai Airport lines, as well as the Guangfo Ring Line. This aims to build a GBA rail transit operation model characterized by "One Network, One Ticket, String of Cities." In the future, Guangzhou Metro's intelligent technologies and service models will radiate throughout the GBA, promoting the integrated development of regional rail transit.

Regarding cross-system interconnection, Guangzhou Metro is advancing the "Four-Network Integration" – the seamless integration of high-speed rail trunk lines, intercity railways, regional railways, and the urban metro network. Intelligent technologies will facilitate seamless transfers between different types of rail systems. For instance, the tag-based fare gates implemented by PDT at Pazhou Station already enable convenient interchange within the paid area between intercity rail and metro services. In the future, passengers might need only "One Card/One Code" to travel seamlessly across these intercity rail and metro systems.

16.5.4 Green and Sustainable Development

Green and low-carbon operation is a crucial direction for the intelligent metro. Guangzhou Metro has made progress in intelligent environmental control and energy saving. The system based on hybrid models and AI algorithms, utilizing Digital Twin and deep learning technologies, optimizes the Environmental Control System (ECS) in metro stations in real-time, tapping into deep energy-saving potential. In the future, through smart energy management systems, metro stations will achieve coordinated optimization of energy production, storage, and consumption, significantly reducing operational energy consumption and carbon emissions.

16.5.5 Challenges and Countermeasures

The intelligent development of Guangzhou Metro still faces numerous challenges. These include the high complexity of technology integration, where compatibility and interoperability between different systems and vendor equipment remain technical hurdles; increased pressure concerning data security and privacy protection, especially with the widespread application of biometric data like facial recognition, making data security and passenger privacy paramount concerns; and the difficulty in balancing investment and benefits, as intelligent upgrades require substantial capital investment, yet their benefits are often difficult to quantify directly.

To address these challenges, Guangzhou Metro is adopting a series of countermeasures: reducing system integration complexity through standardization and open platforms; establishing a multi-layered security protection system to safeguard data security; and balancing investment and benefits through whole-life-cycle cost assessments and innovative business models. These measures will assist Guangzhou Metro in overcoming obstacles in its intelligentization journey and achieving sustainable development.

16.6 Building an International Vocational Education System for the Intelligent Metro

During its intelligent metro development, Guangzhou Metro has accumulated rich technical experience and management methodologies. Transforming this knowledge into vocational education resources and constructing an international vocational education system represents a crucial pathway for promoting intelligent metro technologies and standards globally.

16.6.1 Curriculum System Architecture

The international vocational education system for the intelligent metro should be competency-based, with curriculum content designed across a four-tiered structure: "Fundamental Theory – Technical Skills – Practical Application – International Perspective."

- **Fundamental Theory Module:** Covers the architecture of intelligent metro systems, principles of key technologies, and relevant standards and specifications.
- **Technical Skills Module:** Focuses on specialized skills such as SuiTeng OS platform operation, maintenance of intelligent equipment, and data analysis.
- **Practical Application Module:** Cultivates trainees' ability to solve practical

problems through real-world case studies and scenarios.

- **International Perspective Module: Introduces** international standards, global technology trends, and cross-cultural communication skills.

Curriculum development should fully incorporate Guangzhou Metro's practical experience. For instance, successful cases like the development and application of SuiTeng OS, the tunnel comprehensive detection system, and intelligent customer service robots should be integrated into teaching materials. Simultaneously, drawing on Guangzhou Metro's experience in managing intercity railways and overseas projects, adaptive courses tailored to different cultural backgrounds and regional characteristics should be developed.

16.6.2 Training Mode Innovation

Vocational education for the intelligent metro should innovate its training methods, adopting a blended model combining virtual and physical environments, and alternating between study and work practice. Virtual simulation technology should be utilized to create virtual training environments for intelligent maintenance, passenger flow guidance, emergency command, etc., allowing trainees to conduct hands-on practice in simulated scenarios. Learning from the experience of Guangzhou Metro's "National Engineering Laboratory," intelligent metro training bases should be established, providing real equipment operation and system debugging environments.

A cultivation model featuring "University-Enterprise Cooperation, Work-Study Integration" should be promoted, collaborating with universities and vocational colleges to conduct targeted talent training. Guangzhou Metro has already accumulated substantial experience in external training, having provided training services to over 40 domestic and international cities, with a total external training volume exceeding 700,000 person-days. This experience can serve as a valuable reference for the international vocational education system.

16.6.3 Resource Platform Development

An intelligent metro vocational education resource platform should be established, integrating course resources, training software, expert databases, and other content, supporting online and mobile learning. Based on the open-platform concept of SuiTeng OS, a scalable and iterative architecture for the vocational education platform should be constructed, supporting multiple terminal access and personalized learning path planning.

Multilingual teaching resources, including textbooks, videos, and virtual simulation software, should be developed to meet the needs of trainees from different countries. Digital twin technology should be fully leveraged to develop 3D visual teaching models of metro systems, intuitively displaying equipment structure, working principles, and operation and maintenance processes.

Concluding Remarks

Guangzhou Metro's intelligent development journey provides valuable experience for global urban rail transit. Progressing from technology introduction to independent innovation, from single-point breakthroughs to system integration, and from local service to regional Bay Area collaboration, Guangzhou Metro has charted a unique path in intelligent metro development. By constructing a systematic international vocational education system, this experience can be applied and promoted more widely, catalyzing the intelligent development of urban rail transit worldwide. As technology continues to advance and operational concepts keep innovating, the intelligent metro will encounter even broader development prospects, injecting new momentum into the high-quality development of urban rail transit. This educational framework is pivotal for disseminating best practices, building global capacity, and ensuring the sustainable and scalable adoption of intelligent metro solutions.

Chapter 17: The "Five Rams" Robot Matrix and Applications

17.1 "Yang Xiao Ying" (Welcoming Lamb) Customer Service Robot

"Yang Xiao Ying" is the customer service robot within Guangzhou Metro's intelligent robot matrix. Its design purpose is to break through the spatial limitations of traditional service counters, extending proactive services to various corners of the station and realizing an intelligent transformation towards "service finding people."

17.1.1 Technical Configuration and Functional Characteristics

The robot is equipped with a 19-inch high-definition interactive display screen and integrates an advanced voice guidance system and a real-time speech interaction engine. Its primary functions encompass:

(1) **Intelligent Inquiry and Guidance: Capable** of providing passengers with voice-based inquiry and guidance services regarding metro lines, transfer routes, security check procedures, and civilized boarding etiquette.

(2) **Proactive Service Extension: Leveraging** its autonomous mobility, the robot dynamically extends the functionality of traditional fixed service counters into the station concourse areas, enabling close-range interaction with passengers. This allows it to actively approach passengers who may appear to need assistance.

17.1.2 Application Value

The deployment of "Yang Xiao Ying" signifies a paradigm shift in metro passenger services, moving from a "reactive response" model to a "proactive greeting" approach. This transformation effectively enhances service accessibility, reduces waiting times, and improves overall service efficiency and passenger satisfaction.

17.1.3 Technical Parameters and Vocational Training Focus

Key Technical Parameters (Illustrative): Includes dimensions/weight for maneuverability, battery life/charging method for operational continuity, navigation system (e.g., SLAM), sensor suite (e.g., LiDAR, cameras), processing unit, and operating software. Specific payload capacity might also be relevant.

Vocational Training Focus: Training programs should cover robot operation, basic troubleshooting, routine maintenance procedures, utilizing the robot's interface for information retrieval, and understanding its integration within the broader station service ecosystem to facilitate effective human-robot collaboration.

Table 17-1 Technical Specifications and Vocational Training Focus of the "Yang Xiao Ying" Customer Service Robot

Technical Dimension	Parameter Details	Vocational Education Focus
Interaction System	19-inch high-definition interactive display; real-time speech interaction system utilizing a multi-microphone array and noise cancellation technology.	Fundamentals of automatic speech recognition (ASR) and natural language processing (NLP); principles of human-computer interaction (HCI) design for public service robots.
Navigation Method	Autonomous navigation system based on LiDAR SLAM (Simultaneous Localization and Mapping), enabling real-time path planning and dynamic obstacle avoidance.	Principles of robot path planning algorithms; operational characteristics and application scenarios of LiDAR and other environmental perception sensors.
Service Coverage	Dynamic mobile service provision across the entire station concourse area, capable of autonomous patrols and targeted passenger approach based on predefined logic or remote dispatch.	Design and optimization of passenger service workflows; protocols for identifying and handling atypical situations or passenger requests beyond standard scripts.

Core Functions	Provision of journey guidance, real-time information queries, and security check directions; potential integration for handling basic multi-language inquiries.	Foundational knowledge of metro network operations, fare systems, and safety regulations; development of multi-language communication skills and cultural sensitivity for passenger interaction.

17.2 "Yang Xiao Yun" (Transporting Lamb)(Luggage Transport Robot

"Yang Xiao Yun," the luggage transport robot, is specifically designed to address the travel difficulties faced by passengers, particularly those carrying large luggage items or bulky packages, during movement within the station.

17.2.1 Core Technologies

Its core navigation technology is SLAM (Simultaneous Localization and Mapping), integrated with multi-sensor fusion technology. This combination ensures the robot's operational capability within complex and dynamic station environments. Specific technical characteristics include:

(1) Autonomous Path Planning: Capable of autonomously calculating the optimal route to a designated destination based on a pre-loaded station map and real-time environmental data.

(2) Dynamic Obstacle Avoidance: Utilizes its sensor suite (e.g., LiDAR, cameras) to perceive the environment in real-time during movement, enabling intelligent avoidance of both static and dynamic obstacles, including pedestrians, thereby ensuring operational safety.

17.2.2 Service Model and Application Significance

Notably, during the pilot phase at Guangzhou Baiyun Station, passengers in need could request the use of the "Yang Xiao Yun" luggage transport service free of charge by contacting station staff. This service directly enhances the travel experience for passengers carrying heavy items, tangibly reflecting Guangzhou Metro's commitment to its human-centric service philosophy. It alleviates physical strain, improves passenger flow efficiency in concourse areas, and demonstrates the practical application of robotics to solve genuine passenger challenges.

17.2.3 Technical Parameters and Vocational Training Focus

Table 17-2 Technical Parameters and Vocational Training Focus of the "Yang Xiao Yun" Luggage Transport Robot

Technical Dimension	**Parameter Details**	**Vocational Education Focus**
Navigation Technology	Utilizes SLAM (Simultaneous Localization and Mapping) for real-time positioning and pathfinding; employs multi-sensor fusion (e.g., integrating LiDAR, depth cameras, ultrasonic sensors) for robust environmental perception.	In-depth understanding of SLAM technology principles; methods for fusing and interpreting data from disparate sensors; troubleshooting common navigation and perception issues.
Load Capacity	Designed for transporting large luggage items and packages, with specified maximum weight and volume capacities. Robust structural design ensures stability during operation.	Fundamentals of robot structural mechanics related to load-bearing; principles and practice of calculating safe operational loads; importance of adhering to load limits for equipment longevity and safety.
Obstacle Avoidance Capability	Features intelligent obstacle avoidance using real-time data from perception sensors. Algorithms distinguish between static and	Principles of obstacle detection and classification algorithms; functionality and operational limits of the emergency braking system; strategies for predicting pedestrian movement in crowded

	dynamic obstacles, triggering appropriate maneuvers or emergency braking via a dedicated safety system.	environments.
Service Mode	Operates primarily via staff-assisted deployment. Station personnel manage service requests, initiate transport tasks, perform safety checks, and handle the interface with passengers, ensuring smooth service delivery.	Standardized workflows for human-robot collaboration during luggage handling; service handover protocols between staff and robot; effective passenger communication and managing service expectations.

17.3 "Yang Xiao Xun" (Patrolling Lamb) Security Robot

"Yang Xiao Xun" is positioned as a patrol service and security robot, serving as a vital component in Guangzhou Metro's integrated "ground-air" emergency response system and intelligent security network.

17.3.1 Technical Configuration and Perception Capabilities

The robot is equipped with a LiDAR navigation system, providing it with high-precision positioning and environmental modeling capabilities that support all-weather dynamic patrols. At the perception level, "Yang Xiao Xun" is outfitted with multiple sensors:

(1) Dual "Visible Light + Thermal Imaging" Cameras: Enable monitoring under normal lighting conditions while also detecting temperature anomalies through thermal imaging. This allows for the early identification of potential fire hazards or the location of individuals in specific scenarios, such as low-visibility environments.

(2) Smoke Detection System: Provides early warning for fire incidents by detecting particulate matter associated with smoke.

(3) AI-Powered Visual Algorithms: Accurately identify various anomalies, such as water accumulation on floors or a passenger who has fallen, enabling prompt alerts and potential intervention.

17.3.2 Collaborative Work Mechanism and Application Value

"Yang Xiao Xun" does not operate in isolation. It is integrated into the station's operational fabric, capable of relaying identified anomalies in real-time to the Station Control Room. This establishes a new, closed-loop management model of "Dynamic Patrol —Intelligent Early Warning —Timely Response." This model transforms traditional security from passive monitoring to proactive warning and rapid response, significantly enhancing the efficiency and safety of metro stations in handling unexpected incidents.

17.3.3 Technical Parameters and Vocational Training Focus

Table 17-3 Technical Parameters and Vocational Training Focus of the "Yang Xiao Xun" Security Robot

Technical Dimension	Parameter Details	Vocational Training Focus
Perception System	Multi-sensor configuration including LiDAR navigation system, dual visible-light & thermal imaging cameras, and integrated smoke detection sensors.	Principles and methods of multi-sensor data fusion; fundamentals of computer vision and thermal image interpretation; calibration and maintenance procedures for diverse sensor types.
Detection Capability	AI-powered identification of anomalies such as water accumulation, smoke, passenger falls, and other predefined unusual events or behaviors.	Understanding algorithms for anomaly and behavior recognition; core principles of pattern recognition; techniques for validating and minimizing false positives/negatives in detection alerts.
Response Mechanism	Integrated real-time data link with the Station Control Room, enabling immediate alert transmission and integration into the station's incident command workflow.	Mastery of established emergency response protocols; operation of integrated control systems for initiating coordinated responses; effective communication procedures during incident escalation.
Operational Mode	Designed for sustained all-weather dynamic patrol capability, supported by automated	Strategies for effective robot energy/power management; principles

	energy management and self-charging functionality for continuous operation.	and maintenance of autonomous charging technology; planning and optimizing patrol routes and schedules for maximum coverage and efficiency.

17.4 System Collaboration of the Robot Matrix and Implications for Vocational Education

The significance of Guangzhou Metro's "Five Rams" robot matrix extends beyond the intelligence of individual devices, lying more profoundly in their deep integration and synergistic collaboration within the intelligent station system.

17.4.1 System Collaboration and Integration with the Intelligent Station

Guangzhou Metro is actively exploring and establishing deep integration mechanisms between these robotic devices and its intelligent platform. The goal is to construct a synergistic operational model characterized by "Intelligent Detection — Precise Guidance — Efficient Execution." Illustrative examples include:

When "Yang Xiao Xun" (Patrolling Lamb) detects spilled liquid on the floor, it can automatically notify "Yang Xiao Jie" (Cleaning Lamb) to proceed to the location for cleanup.

If "Yang Xiao Fei" (Flying Lamb) monitors passenger flow congestion at an entrance/exit, it can guide "Yang Xiao Ying" (Welcoming Lamb) to the area to make broadcasts.

Operational data from all robots converges into the intelligent platform, where it is utilized for big data analytics to continuously optimize operational strategies, maintenance schedules, and resource allocation.

17.4.2 Implications for International Vocational Education

The application of the "Five Rams" robot matrix presents new requirements for international vocational education in the rail transit sector, while simultaneously providing vivid, real-world teaching cases.

(1) **New Skill Modules: Vocational** curricula must incorporate new skill

modules addressing the operation, maintenance, and management of service robots. These should cover areas such as Service Robot Operation and Maintenance, Human-Robot Collaborative Management, and Robot Dispatch System Operation.

(2) **Practical Teaching Platform: These** robots serve as excellent comprehensive training platforms for intelligent station operation and management. They allow students to develop system-level thinking by engaging with complex, integrated station management scenarios that involve multiple interacting systems.

(3) **Standards and Protocols: As** robot applications become widespread in rail transit, related operational protocols, safety standards, and emergency procedures specific to human-robot environments will become essential components of vocational education.

Concluding Remarks

The pilot application of the "Five Rams" intelligent robot matrix at Guangzhou Baiyun Station represents a cutting-edge exploration within the rail transit industry in the realms of smart services and intelligent maintenance. Through the functional complementarity and systemic collaboration of the three profiled robots – "Yang Xiao Ying," "Yang Xiao Yun," and "Yang Xiao Xun" – Guangzhou Metro is committed to creating a "Panoramic Smart Service Station." This model envisions all-scenario services, whole-process coverage, and system-wide intelligence. This practice not only delivers a smarter and more convenient travel experience for passengers and introduces more efficient management modes for station operations but also provides invaluable "Guangzhou Experience" for the global intelligent upgrade of rail transit and the evolution of international vocational education.

Chapter 18: Application of Big Data and Artificial Intelligence in Operation and Maintenance

18.1 Intelligent Maintenance System Architecture and Technological Foundation

Guangzhou Metro's intelligent maintenance system is fundamentally data-driven. Leveraging advanced technologies such as the Internet of Things (IoT), cloud computing, and artificial intelligence (AI), it has achieved a fundamental paradigm shift from "preventive maintenance" to "predictive maintenance." The system architecture adopts a "Cloud-Edge-Device" collaborative model, establishing an intelligent maintenance ecosystem that covers the entire metro network.

18.1.1 Intelligent Maintenance Platform Architecture

The Guangzhou Metro intelligent maintenance platform employs a four-layer architecture design:

- **Perception Layer: Responsible** for data acquisition. It utilizes sensors and detection devices deployed at critical locations on rolling stock, tracks, tunnels, and other key infrastructure to collect real-time status data.
- **Network Layer: Ensures** reliable data transmission. It comprehensively utilizes communication technologies like 5G and EUHT (Enhanced Ultra-High Throughput) to guarantee low-latency, high-reliability data transfer from the edge to

the cloud.

• **Platform Layer: Provides** data storage, processing, and analytics capabilities. It constructs a big data platform based on cloud computing to integrate and manage multi-source, heterogeneous data.

• **Application Layer: Addresses** specific maintenance scenarios. It develops intelligent analysis and decision support systems for applications such as condition assessment, fault prediction, and maintenance scheduling.

The core of this architecture is the Tencent OS (Intelligent Rail Transit Operating System). This system represents the industry's first new-generation intelligent rail transit OS based on the Industrial Internet of Things (IIoT). Suntec OS adopts an innovative, open, and component-based design philosophy, integrating key technologies from industrial control, the internet, IoT, and cloud computing. It connects downwards to massive industrial equipment and systems and upwards supports the rapid development and iteration of intelligent applications.

18.1.2 Data Acquisition and Processing System

Guangzhou Metro has established a comprehensive data acquisition system covering major subsystems including rolling stock, permanent way, power supply, and communication & signaling. For data acquisition, various types of sensors are deployed—including vibration sensors, temperature sensors, and high-definition image sensors—enabling real-time monitoring of the status of critical equipment. Taking the Line 10 trains as an example, their intelligent maintenance system can monitor the status of key train components in real-time and instantly transmit this status data to the core platform – the Tencent OS.

Regarding data processing, Guangzhou Metro utilizes its big data platform to integrate, store, and manage multi-source heterogeneous data. After intelligent analysis by the platform, potential issues are identified, and early warnings are generated. Furthermore, the platform automates train maintenance scheduling and work order dispatch, providing robust technical support for the safe and smooth operation of train services. The data-driven intelligent decision-making system, through the analysis of full life-cycle vehicle data, enables the simulation, evaluation, and optimization of the entire vehicle operation process. This facilitates more informed strategic planning for maintenance and resource allocation.

Table 18-1 Data Types and Acquisition Methods in Guangzhou Metro's Intelligent Maintenance System

Data Category	Acquisition Method	Acquisition Frequency	Application Scenario
Vehicle Status Data	On-board sensors (vibration, temperature, etc.); 360-degree image inspection systems utilizing a multi-camera setup and computer vision.	Real-time / Near-real-time stream	Enables fault diagnosisand forms the foundation for predictive maintenance models for critical train components.
Tunnel Structure Data	Dedicated tunnel inspection vehicles equipped with integrated systems (e.g., LiDAR, high-res imaging); periodic manual surveys.	Periodic (e.g., Monthly / Quarterly cycles)	Supports comprehensive tunnel health assessment, monitoring for structural defects, deformation, and lining degradation.
Track Geometry Parameters	Track inspection vehicles (e.g., track geometry cars); fixed or mobile intelligent inspection devices using laser and inertial measurement.	Periodic (e.g., Weekly / Monthly schedules)	Informs maintenance decisions for track alignment and smoothness adjustment, ensuring ride quality and safety standards.
Passenger Flow Data	Automated fare gate transaction logs; video analytics from station surveillance cameras; Wi-Fi/Bluetooth probing from passenger mobile devices (anonymized).	Continuous real-time aggregation	Powers passenger flow prediction models and enables dynamic crowd management and guidance strategies.
Environmental	Network of fixed sensors	Continuous real-time	Facilitates proactive

Data	monitoring temperature, humidity, smoke/particulates, and water leakage at critical locations.	monitoring	environmental monitoring and triggers automated alerts for emergency response to hazards like fire or flooding.

18.2 Innovative Applications of Key Intelligent Technologies in Operation and Maintenance

18.2.1 Intelligent Tunnel Inspection Technology

The tunnel comprehensive detection system developed by Guangzhou Metro represents a significant technological breakthrough in the field of tunnel inspection. This system achieves sub-millimeter level tunnel detection while operating at speeds of 60 km/h, improving efficiency by approximately 120 times compared to manual inspection. This capability signifies that Guangzhou Metro has attained rapid and precise tunnel inspection capabilities, marking the entry of tunnel inspection into an intelligent phase.

The system integrates technologies including LiDAR scanning, 3D structured light measurement, and high-definition line-scan camera imaging. It can accurately capture defects such as tunnel cracks, spalling, water leakage, and segment misalignment, achieving a maximum accuracy of 0.3 mm. The metro operation project team overcame challenges related to the complex tunnel environment and the high precision requirements for multi-equipment coordination. This involved over 30 scheme optimizations and more than 100 algorithm iterations, ultimately successfully breaking through the bottlenecks associated with vehicle mounting and system debugging. A companion Network-Rail-Tunnel data comprehensive analysis platform was also developed, possessing 3D visualization processing capabilities for the inspection data.

As a case of technological breakthrough within China's rail transit industry, this system integrates deep learning algorithms, digital twin technology, and other advanced methods to achieve tunnel health monitoring. A relevant 负责人 from

Guangzhou Metro stated that, based on the trial operation results and aligned with business needs, they will continue to expand and develop the system's functionalities, providing greater technical support for the safe operation of the Guangzhou Metro network.

18.2.2 Intelligent Vehicle Inspection System

Guangzhou Metro has achieved a transformation in the field of vehicle maintenance, shifting from traditional manual inspection to an intelligent "human-machine combined" maintenance model. The industry's first urban rail vehicle intelligent inspection robot utilizes advanced technologies including intelligent robotics, SLAM autonomous navigation, multi-degree-of-freedom flexible collaborative robotic arm control, high-definition optical imaging, image pattern recognition, and AI deep learning algorithms.

This intelligent inspection robot can perform 2D/3D high-definition imaging of key inspection points underneath the train car body, intelligently judge the abnormal status of key undercarriage components, and achieve automatic transfer across multiple tracks. It also automates the processes of vehicle inspection, intelligent analysis, and fault diagnosis. Furthermore, it can adapt to multi-robot collaborative operations, addressing the shortcomings of existing metro vehicle inspection equipment, such as the inability to automatically switch tracks and the presence of inspection blind spots. Since its trial use began in April 2020 at the Qishan Depot on Line 9, this robot has cumulatively inspected 2,241 trains, achieving a fault recognition rate of 98.1%.

Working in synergy with the intelligent inspection robot is the 360° train image online detection system. This system employs high-definition line-scan technology with near-infrared laser fill-light, image recognition technology, and deep learning technology. It covers the automated detection and identification of visually inspectable key components across the entire 360° range of the train, including the roof, sides, and undercarriage. Fixed installations of the 360° system around the tracks can scan and inspect components across the entire "body" of a moving train, covering over 9,000 inspection items. Conversely, the intelligent inspection robot is responsible for performing precise inspections on key components of stationary trains, covering 2,478 inspection items.

18.2.3 Electronic Component Health Prediction Management

Within Guangzhou Metro's data-driven Prognostics and Health Management (PHM) system, the degradation detection and health management of urban rail vehicle electronic boards is a crucial application area. Using data-driven PHM methodologies, Guangzhou Metro has conducted failure mechanism analysis and lifetime prediction assessments for electronic components widely used on these boards that exhibit high failure rates, such as electrolytic capacitors, film resistors, and MOSFETs.

Applying the data-driven PHM approach, combined with a case study involving a vehicle braking system power supply board, Guangzhou Metro has achieved significant cost reduction and efficiency improvement in practice. This method emphasizes condition monitoring and data collection during the early phases. By collating and summarizing this data alongside practical operational experience, a full life-cycle health prediction model for electronic boards has been constructed. This model enables more proactive maintenance planning and resource allocation for these critical components.

18.3 Intelligent Analysis, Early Warning, and Decision Support Systems

18.3.1 Big Data-Driven Fault Prediction

Leveraging big data and cloud computing technologies, Guangzhou Metro has constructed an intelligent maintenance system for metro vehicles. This system facilitates the simulation, evaluation, and optimization of the entire vehicle operation process by establishing a robust intelligent decision-making system, enhancing work coordination, and enabling information sharing. Founded on full life-cycle vehicle data, the system provides comprehensive management of faults and component lifespan throughout the vehicle's operational life.

In the domain of fault prediction, the system analyzes historical maintenance data, real-time operational data, and environmental data to establish fault prediction models. This enables an evolution from "reactive maintenance after failure" to "preventive maintenance" and further towards "predictive maintenance." The "Jia Yi Wei" intelligent maintenance system, developed by PDT Science & Technology and piloted on the second phases of Lines 13 and 14, exemplifies this approach. It performs 7x24-hour "precise condition monitoring" of equipment and facilities, achieving fault warnings and accurate condition assessment, thereby fundamentally revolutionizing the traditional maintenance model.

18.3.2 Intelligent Passenger Flow Prediction and Management

In passenger flow management, Guangzhou Metro has proposed a Gated Convolutional Graph Neural Network (GCGRU) model for urban rail transit passenger flow prediction. This model first analyzes the spatial relationships of short-term passenger flows within the urban rail network, utilizing a Graph Convolutional Network (GCN) to extract the spatial interactions between passenger flows at different stations. Subsequently, it analyzes the temporal evolution of passenger flows at each station in the network, employing a Gated Recurrent Unit (GRU) to capture the temporal characteristics of the passenger flow data at each station. This integrated approach forms a data-driven model for short-term passenger flow prediction across the urban rail transit network.

Compared to traditional methods, this model more effectively characterizes the spatiotemporal correlations of passenger flows across the network's stations and deeply explores the underlying mechanisms of passenger flow variations. Furthermore, compared to existing GCN models, this model introduces a travel-time-oriented adjacency matrix, enabling it to uncover the intrinsic relationship between passenger flow data and train schedule data, resulting in higher accuracy and improved interpretability. Validation results, using the prediction of exit passenger volume at typical Guangzhou Metro stations as an example, demonstrate that this model outperforms existing models in both overall prediction performance and the predictive performance at individual stations.

On Lines 10 and 12, PDT Science & Technology has adopted innovative technologies such as "Multi-modal Fusion Passenger Flow Prediction" and "Dynamic Crowd Control Intelligent Decision-making." These provide solutions for the

AI-driven governance challenges of large passenger flow hub stations. They assist in achieving precise passenger flow management driven by Big Data + AI, offer core technical support for alleviating the passenger flow pressure on Line 3, and enable comprehensive dynamic early warning, precise passenger guidance, and transport capacity optimization across the stations.

18.4 Implementation Effectiveness and Typical Cases of Intelligent Maintenance

18.4.1 Maintenance Model Transformation and Efficiency Improvement

Guangzhou Metro boldly initiated a reform of its vehicle maintenance model on Line 9, utilizing the "Intelligent Inspection Robot + 360° Train Online Detection System" to perform automated vehicle inspections. This marked the beginning of a "human-machine combined" vehicle inspection paradigm. This new model addresses inherent limitations of manual operations, such as susceptibility to fatigue, limited efficiency, and inconsistency, thereby further ensuring the safety and quality of vehicle maintenance tasks. Concurrently, it provides technical support for optimizing operational scheduling and shift patterns.

Under this new model, the number of maintenance personnel required for night shifts at the Qishan Depot on Line 9 has decreased by 37.5% compared to the traditional manual operation mode. This allows maintenance staff to perform more inspection and maintenance work during daytime hours, reducing the labor intensity associated with night shifts. This transformation not only enhances work efficiency but also improves the overall work experience for employees.

18.4.2 Intelligent Image Interpretation and Security Check Efficiency Enhancement

With the application of the SuiTeng OS, a significant amount of mechanical and repetitive work in metro operation management can be reduced. Cai Changjun, Deputy General Manager of Guangzhou Metro Group, provided an illustrative example: "For instance, Guangzhou Metro has seven to eight hundred security checkpoints. Each checkpoint previously required an image interpretation officer. After implementing AI-powered intelligent image interpretation for security checks, the number of dedicated image interpreters has been reduced from 2,400 to just over

70. This not only improves the efficiency and accuracy of image interpretation but also drastically reduces security screening costs."

The application of the AI-based image interpretation system has not only substantially lowered labor costs but also enhanced the accuracy and efficiency of threat detection. Utilizing deep learning algorithms, the system can rapidly identify prohibited items within luggage, reducing subjectivity in human judgment and misjudgments caused by fatigue, thereby further elevating the safety standards of metro security checks.

18.4.3 Construction of the Panoramic Smart Service Station

Guangzhou Metro piloted the application of an intelligent robot matrix at Guangzhou Baiyun Station, involving the concentrated deployment of intelligent service robots covering five primary functions: passenger service, luggage transport, patrol service, cleaning, and aerial inspection – namely, "Yang Xiao Ying," "Yang Xiao Yun," "Yang Xiao Xun," "Yang Xiao Jie," and "Yang Xiao Fei."

These robots perform distinct roles:

"Yang Xiao Xun" (Patrolling Lamb), the patrol service robot, equipped with a LiDAR navigation system, conducts all-weather dynamic inspections of equipment, facilities, and passenger flow. Fitted with dual "visible light + thermal imaging" cameras and a smoke detection system, it accurately identifies abnormalities such as water accumulation, smoke, or passenger falls, relaying this information in real-time to the station control room.

"Yang Xiao Fei" (Flying Lamb),the aerial inspection robot, employs "low-altitude + 3D" technology, carrying high-precision LiDAR and a thermal imaging gimbal. Using 3D modeling and panoramic aerial photography, it performs real-time video capture and dynamic flood monitoring at station entrances/exits, thereby enhancing the metro's integrated "ground-air" emergency response system layout.

Guangzhou Metro plans to debug and upgrade the robots based on the pilot results at Guangzhou Baiyun Station and in line with the specific demands of the metro environment. This involves continuously refining service details and expanding the breadth and depth of applications. The aim is to provide passengers with a smarter and more convenient travel experience, deliver more efficient management for station operations, and ultimately establish Guangzhou Baiyun Station as a benchmark "Panoramic Smart Service Station."

Table 18-2 Implementation Performance Statistics of Guangzhou Metro's Intelligent Maintenance System

Application System	**Implementation Site / Line**	**Efficiency Improvement**	**Labor Savings**	**Fault Recognition Rate / Key Metric**
Tunnel Comprehensive Detection System	Pilot deployment across network	Inspection efficiency improved by ~120x compared to manual methods	Data not systematically available	Achieves sub-millimeter accuracy (0.3mm) in defect detection
Intelligent Inspection Robot	Qishan Depot, Line 9	Enabled significant operational shift; facilitated 37.5% reduction in night-shift personnel	Significant reduction in manual inspection workload	Demonstrates a high 98.1% fault recognition rate in practice
360° Train Image Detection System	Line 9	Capable of automated inspection covering over 9,000 visual checkpoints per train	Reduces manual visual inspection burden	Data not systematically available; focuses on comprehensive visual coverage
AI-Based Image Interpretation System	Network-wide implementation (Security Checkpoints)	Dramatically increased processing throughput; enabled massive personnel reduction	Substantial savings, reducing dedicated staff from ~2,400 to ~70	Reported increase in detection accuracy and consistency

Intelligent Robot Matrix	Guangzhou Baiyun Station (Pilot)	Enables 24/7 service availability across multiple functions (guidance, patrol, cleaning, etc.)	Partial replacement of human labor for specific repetitive or hazardous tasks	Capable of identifying multiple anomaly types (e.g., spills, congestion, unauthorized access)

18.5 Implications of Intelligent Maintenance Technologies for International Vocational Education

The practical experience gained by Guangzhou Metro in applying big data and artificial intelligence to operation and maintenance holds significant reference value for international vocational education within the rail transit sector. These technological applications not only enhance maintenance efficiency but also provide new directions and content for talent cultivation.

18.5.1 Updating the Vocational Education Curriculum System

The development of intelligent maintenance technologies necessitates updates to the vocational education curriculum for rail transit. New course modules need to be added, covering areas such as Intelligent Detection Technologies, Big Data Analytics, Artificial Intelligence Algorithms, and Robot Operation and Maintenance. These aim to equip students with the skills to operate and maintain intelligent maintenance systems and perform data analysis.

Taking Guangzhou Metro's Tunnel Comprehensive Detection System as an example, vocational education must encompass the principles and applications of technologies like LiDAR scanning, 3D structured light measurement, and high-definition line-scan camera imaging, alongside methods for analyzing and processing the resulting inspection data. For instruction on intelligent inspection robots, content should include SLAM autonomous navigation technology, control techniques for multi-degree-of-freedom flexible collaborative robotic arms, and image pattern recognition technology.

18.5.2 Practical Training and Skills Development

Guangzhou Metro's intelligent maintenance practices provide rich material for practical teaching cases in rail transit vocational education. By establishing training platforms that replicate real-world maintenance environments, students can gain hands-on experience operating intelligent detection equipment, analyzing operational data, and performing fault diagnosis and prediction. This cultivates practical operational abilities and problem-solving skills.

Specifically, practical teaching projects can be designed around areas such as:

- Operation and maintenance of intelligent inspection robots.
- Analysis and processing of tunnel inspection data.
- Construction and optimization of passenger flow prediction models.

Through this project-driven teaching approach, students can gain a deep understanding of the application scenarios and operational workflows of intelligent maintenance technologies, laying a solid foundation for future careers in intelligent rail transit maintenance.

18.5.3 International Certification and Standards Alignment

Guangzhou Metro's practical application of intelligent maintenance technologies provides a technical foundation and standard reference for the international certification of rail transit vocational education. By integrating Guangzhou Metro's technical standards, operational procedures, and personnel training requirements into vocational curricula, the cultivation of talent can be better aligned with international practices, thereby enhancing the global relevance of the vocational education provided.

Furthermore, when Guangzhou Metro's intelligent maintenance technologies are exported to international markets, they require corresponding support from technically skilled personnel. Through collaboration with international vocational education institutions, developing training courses and certification systems based on Guangzhou Metro's intelligent maintenance practices can help cultivate technical talents suited to the demands of the global market. This, in turn, promotes the international dissemination of China's rail transit technologies.

Chapter Summary

Guangzhou Metro's practices in applying big data and AI to operation and maintenance represent the current forefront of intelligent maintenance in urban rail transit. Through the construction and application of systems like the Tunnel Comprehensive Detection System, the Intelligent Vehicle Inspection System, Electronic Component Health Prediction Management, and the Intelligent Passenger Flow Prediction and Management System, Guangzhou Metro has achieved significant improvements in maintenance efficiency and effective reductions in operational costs. This provides valuable, transferable experience for the intelligent development of urban rail transit globally.

These technological innovations and practical experiences not only provide robust support for the high-quality development of Guangzhou Metro itself but also offer a valuable reference for the intelligent upgrade of urban rail transit systems worldwide. By establishing a systematic international vocational education system, these innovations and experiences can be applied and promoted more broadly, propelling the intelligent evolution of operation and maintenance technologies in the global urban rail transit sector. As technology continues to advance and application scenarios expand, the use of big data and AI in Guangzhou Metro's maintenance operations will deepen further, offering even stronger technical guarantees for the safe, efficient, and green operation of urban rail transit.

Chapter 19: Application of Digital Twin and AR/VR Technologies in Metro Systems

19.1 Digital Twin Technology Overview and System Architecture

Digital Twin, as a key technology for realizing the interconnection and interoperability between the physical and information worlds, plays a significant role throughout the entire lifecycle of Guangzhou Metro, encompassing planning and design, construction implementation, and operation and maintenance management. Through multi-dimensional virtual modeling, sensor data acquisition, real-time data interaction, and dynamic simulation and deduction, this technology constructs a highly accurate virtual digital mirror that corresponds directly to the physical metro system. It provides unprecedented digital management capabilities for the metro infrastructure.

19.1.1 Digital Twin Technology Framework

The Guangzhou Metro Digital Twin system adopts a five-layer architecture design, comprising from bottom to top:

- **Physical Twin Layer:** Consists of the physical entities, including tracks, trains, station environments, and other infrastructure assets.
- **Modeling & Mapping Layer:** Creates the digital representation, typically through the fusion of Building Information Modeling (BIM) and Geographic Information Systems (GIS) data, ensuring geometric and semantic accuracy.
- **Data Layer:** Manages multi-source, heterogeneous data streams, integrating real-time sensor data (IoT), historical maintenance records, operational schedules, and the static model data.
- **Analysis & Services Layer:** Hosts simulation and analytics engines, such as finite element analysis for structural integrity and multi-body dynamics for train movement simulations, enabling predictive capabilities and performance evaluation.
- **Application & Services Layer:** Provides 3D visualization interfaces and decision-support tools for end-users, translating data and model outputs into actionable insights for operators and managers.

This architecture achieves a precise mapping from physical entities to their virtual counterparts, forming the foundational technological bedrock for intelligent maintenance and operational optimization.

In Guangzhou Metro's implementation, the Digital Twin platform integrates multi-source data – including BIM models, GIS context, and real-time IoT sensor data – with dynamic simulation technologies like finite element analysis and multi-body dynamics. This integration facilitates comprehensive lifecycle management of rail transit infrastructure. The platform can mirror the status of physical assets in real-time, perform simulation-based analysis and prediction, and thereby provide a scientific basis for operation and maintenance decision-making.

19.1.2 Key Technology Components

The realization of Digital Twin technology relies on the synergy of several key technologies:

- **IoT Sensing Technology:** Provides the data acquisition capability for the Digital Twin. Sensors deployed at critical locations on vehicles, tracks, tunnels, and within stations collect real-time data on equipment status, environmental conditions, and operational parameters.

- **BIM Modeling Technology:** Constructs high-precision 3D models of metro facilities. These models are rich in both geometric detail and non-graphic attribute information, serving as the definitive digital representation of the physical assets.

- **GIS Technology:** Offers spatial and geographic information support. It enables the integration of the metro environment within the broader urban context, supporting spatial analysis and city-scale planning interfaces.

- **Data Integration Technology:** Addresses the challenge of fusing and managing multi-source, heterogeneous data. It ensures that disparate data types from various sources can be consolidated, correlated, and accessed coherently within the Digital Twin environment.

- **Simulation & Analysis Technology:** Employs algorithmic models and computational methods – such as finite element analysis for stress/strain evaluation or AI models for predictive maintenance – to assess current equipment states, predict future behavior, and simulate potential scenarios.

Table 19-1 Digital Twin Technology Components and Functional Description

Technology Type	Primary Function	Application in Guangzhou Metro
IoT Sensing Technology	Enables real-time data acquisition and continuous status monitoring of physical assets and environments through deployed sensor networks.	Deployment of structural health monitoring sensors in tunnels and comprehensive vehicle status monitoring sensors tracking critical subsystems like propulsion, braking, and door operations.
BIM Modeling Technology	Facilitates the creation of detailed 3D digital models of facilities, integrating both geometric data and rich information management for assets throughout	Development and utilization of high-precision BIM models for metro stations and tunnels, serving as the foundational geometric and information core for the Digital Twin platform.

	their lifecycle.	
GIS Technology	Provides robust capabilities for spatial geographic information management and analysis, enabling contextual placement within the broader urban environment.	Applied extensively in new line route planning and optimization, and for conducting analysis of interactions with the surrounding urban environment (e.g., land use, population density).
Data Integration Technology	Solves the critical challenge of fusing and managing multi-source, heterogeneous data (e.g., real-time sensor feeds, static models, maintenance records) into a coherent, accessible data environment.	Implementation of a unified comprehensive operation and maintenance data management platform that ingests, correlates, and manages diverse data streams for holistic system awareness.
Simulation & Analysis Technology	Empowers status assessment of current conditions and predictive analysis of future states or potential scenarios through advanced computational models and algorithms.	Used for sophisticated tunnel structural health assessment and prediction, employing models to simulate long-term deformation, stress distribution, and potential failure modes.

19.2 Innovative Applications of Digital Twin in Metro Operation and Maintenance

19.2.1 Intelligent Tunnel Inspection and Health Monitoring

The tunnel comprehensive detection system developed by Guangzhou Metro is an exemplary application of Digital Twin technology in the maintenance domain. This system achieves sub-millimeter accuracy tunnel inspection while operating at speeds of 60 km/h, improving efficiency by approximately 120 times compared to manual inspection. This capability signifies that Guangzhou Metro has attained rapid and precise tunnel inspection capabilities.

The system integrates technologies including LiDAR scanning, 3D structured light measurement, and high-definition line-scan camera imaging. It can accurately capture defects such as tunnel cracks, spalling, water leakage, and segment misalignment, achieving a maximum accuracy of 0.3 mm. The metro operation project team overcame significant challenges, including the complex tunnel environment and the high precision requirements for multi-equipment coordination. This process involved over 30 scheme optimizations and more than 100 algorithm iterations, ultimately succeeding in breaking through the bottlenecks associated with vehicle mounting and system debugging. A companion Network-Rail-Tunnel data comprehensive analysis platform was developed concurrently, possessing 3D visualization processing capabilities for the inspection data.

This system integrates deep learning algorithms and Digital Twin technology to achieve tunnel health monitoring. By constructing a Digital Twin model of the tunnel structure, the system can perform real-time comparisons between the actual state and the designed state, predict the evolution trends of the tunnel structure, and provide a decision-making basis for preventive maintenance. A relevant 负责人 from Guangzhou Metro stated that, based on the trial operation results and aligned with business needs, they will continue to expand and develop the system's functionalities, providing more robust technical support for the safe operation of the Guangzhou Metro network.

19.2.2 Intelligent Infrastructure Operation and Maintenance Platform

In the construction of Guangzhou Metro Lines 10 and 12, PDT Science & Technology has implemented Digital Twin technology throughout the entire lifecycle of both lines, encompassing construction and subsequent operation and maintenance. This application spans from commissioning control during the construction phase to simulation and decision support during the operational maintenance phase, effectively enhancing project quality and operational efficiency.

To meet the requirements for intelligent operation, passenger service, safety maintenance, and smart emergency response on the new lines, PDT independently developed an intelligent operation management and control platform. This platform equips Lines 10 and 12 with intelligent "Station Brains." It deeply integrates over 20 subsystems from different manufacturers using different protocols, achieving data fusion and cross-equipment coordination. Through this unified platform enabling "one-click linkage," it facilitates intelligent assisted operation for scenarios such as one-click station opening/closing, video patrol inspection, passenger flow early warning, operational dispatch, emergency plan activation, and maintenance assistance.

The WebGIS-based Digital Twin platform provides a full lifecycle management solution for rail transit infrastructure. By integrating BIM, GIS, and IoT sensor data, the platform enables the real-time monitoring and predictive maintenance of track fasteners. Using the Shanghai Metro Longyang Road - Zhangjiang Hi-Tech Park experimental line as a case study, the advantages of this solution in 3D visualization and intelligent decision support were validated. It provides a digital solution facilitating the transition of rail transit maintenance from a reactive model to a predictive one.

19.2.3 Pipeline Management and Conflict Detection

Practices in Chongqing regarding its digital pipeline platform offer valuable reference for Guangzhou Metro. Chongqing has explored the application path of the digital pipeline platform in the planning and design phase of rail transit projects, injecting "intelligent genes" into the process.

The core of this application lies in the deep integration of digital pipeline technology with the planning workflow:

- On one hand, it constructs a high-precision 3D Digital Twin of underground pipelines, integrated with BIM. This allows designers to virtually "see through" the ground, understanding key information such as pipeline location, routing, and burial depth with millimeter-level accuracy.

- On the other hand, designers can intuitively assess the spatial relationship between the proposed rail alignment and existing pipelines. This helps avoid conflicts from the very outset of planning, optimizes the alignment and station design, enables scientific decision-making, and significantly shortens the survey and scheme comparison cycle.

To date, Chongqing has initiated pipeline verification for 15 planned rail transit lines in their preliminary stages, precisely marking 276 key locations. The next step involves advancing more refined "rail-pipeline relationship verification" to optimize designs with millimeter-level precision, eliminating risks at the drawing stage, and avoiding project delays and cost overruns. This methodology holds significant reference value for Guangzhou Metro in managing pipelines during both new line construction and the renovation of existing lines.

19.3 Application of AR/VR Technologies in Metro Services and Operation Maintenance

19.3.1 AR-enabled Smart Inspection and Remote Collaboration

SuiTeng Digital Technology Co., Ltd., established by Guangzhou Metro and Tencent, introduced HiScene as an ecosystem partner in the field of Augmented Reality (AR) to promote the digital and intelligent development of rail transit. HiScene developed a digital-intelligent platform for production operation and maintenance in the rail transit industry, which was selected for the 2023 Pioneer Virtual Reality Application Cases list jointly released by the Ministry of Industry and Information Technology, the Ministry of Education, the Ministry of Culture and Tourism, the National Radio and Television Administration, and the General Administration of Sport of China.

Based on its spatial AI computing platform featuring an integrated device-cloud architecture and combined software-hardware capabilities, and leveraging innovative technologies such as AR wearable spatial computing, large AI models, and big data analytics, HiScene constructed an AR smart inspection system. This system is built around three core applications: "Work Guidance + Intelligent Recognition + Multi-person Collaboration," providing the metro with capabilities including digital-intelligent work orders, digital-intelligent collaboration, and information sharing.

This system deeply integrates AR technology with AI to form a new quality productive force, connecting rail transit with the digital world and using digitalization to boost the development of rail transit and enhance service levels. Maintenance personnel wearing AR devices can access real-time equipment information, operational guidance, and fault alerts, significantly improving inspection efficiency and accuracy. Simultaneously, remote experts can guide on-site personnel through the repair of complex equipment via a remote collaboration system, reducing expert travel time and increasing problem-resolution efficiency.

19.3.2 VR Immersive Experience and Virtual Training

The Guangzhou Metro Museum is a prime example of VR technology application. Focusing on scientific knowledge dissemination, the museum integrates various new technologies and pioneered a digital metro experience space.

Zhizaiyuntian utilized multimedia technologies including VR virtual scenes, Kinect somatosensory interactive games, a giant shaped Tianshu projection screen, OLED transparent interactive touchscreens, multi-touch scenes, and 3D Mapping on a shield machine model to highlight fifteen major features. Here, visitors can not only see the development history of Guangzhou Metro but also experience the working details of metro operation and construction firsthand and appreciate the professional spirit of metro staff. This has enabled Guangzhou Metro to establish a genuine display and interaction platform for public communication, generating significant social impact.

Among these, the metro train front 3D Mapping projection technology projects customized animations onto a physical train front, creating a strong visual impact. The shield machine 3D Mapping projection features the nation's only true-scale shield machine model and imaging system, complemented by an LED screen simulating a construction site overhead, resulting in a fantastical and stunning visual effect using 3D Mapping projection on the shield machine model.

Since its opening, the Metro Museum has become one of the most popular science and technology venues for Guangzhou citizens. It has received 500,000 visitors, with a daily average footfall of 1,000 people, and has hosted over 1,000 group tours. This application of VR technology not only enhances public knowledge about metros but also provides an immersive learning environment for staff training.

19.3.3 AR Real-Scene Navigation and Intelligent Guidance

The multifunctional smart terminal deployed at Shenzhen North Station on MTR (Shenzhen) Line 4 is an innovative product in China's rail transit sector. It possesses AI visual recognition and positioning technology and supports connection to the mobile-based MTR (Shenzhen) Live+ Mini Program, simultaneously providing dual navigation services: AR real-scene navigation and VR virtual navigation.

This product offers passengers multiple services, including intelligent AI voice inquiry, digital interaction, ticket processing, information query, and remote customer service. The multifunctional smart terminal features in-station real-scene navigation. Passengers can select a target within the station or its surroundings by clicking on the scene model within the device, which then intelligently plans a path and plays a route video. Simultaneously, passengers can scan the QR code on the device to connect to the MTR (Shenzhen) Live+ Mini Program to obtain real-time navigation routes.

This dual-port navigation not only provides accessible route planning for passengers in wheelchairs, those carrying large luggage, or pushing strollers but also enables continuous navigation services through the MTR (Shenzhen) Live+ Mini Program, offering precise interaction that is both convenient and fast. The multifunctional smart terminal uses a 55-inch large screen, is simple to operate, and employs UZi, the MTR (Shenzhen) mascot, as a virtual digital service avatar. It answers passenger questions regarding route guidance, ticket handling, station and operational information, travel, and surrounding area recommendations flexibly and rapidly, truly achieving intuitive and barrier-free communication services.

19.4 Implementation Effectiveness of Digital Twin and AR/VR Technologies

19.4.1 Enhancement of Operational and Maintenance Efficiency

The deployment of Guangzhou Metro's tunnel comprehensive detection system has increased tunnel inspection efficiency by approximately 120 times compared to manual methods. Capable of performing sub-millimeter accuracy inspections at speeds of 60 km/h, this system signifies that Guangzhou Metro has acquired rapid, precise tunnel inspection capabilities, marking the entry of tunnel inspection into an intelligent era.

In the field of rolling stock maintenance, the industry's first urban rail vehicle intelligent inspection robot utilizes a suite of advanced technologies, including intelligent robotics, SLAM autonomous navigation, multi-degree-of-freedom flexible collaborative robotic arm control, high-definition optical imaging, image pattern recognition, and AI deep learning algorithms. It can perform 2D/3D high-definition imaging of key inspection points beneath the train chassis and intelligently assess the abnormal status of critical undercarriage components.

Predictive maintenance strategies based on Digital Twins have significantly reduced operational and maintenance costs. Research indicates that achieving early prediction of fastener damage through dynamic simulation, thereby shifting the maintenance model from reactive response to predictive maintenance, can lower O&M costs by 30% and extend asset service life.

19.4.2 Improvement in Service Quality

On Guangzhou Metro Lines 10 and 12, all stations are equipped with the newly upgraded, self-developed smart travel product series from PDT. This provides passengers with intelligent services throughout their journey, including intelligent inquiries and seamless fare payment. Among these, the smart fare gate payment zones incorporate CRcode, NFC, and Face ID technologies, while also adopting new swing gate passage units, creating more convenient and secure passenger access channels.

The new generation upgraded smart customer service centers integrate technologies such as AI voice interaction and biometric recognition. They enable passengers to independently complete various self-service functions, including ticket transactions and information inquiries. Passengers can register their facial features at these centers to activate face-scan access through the Guangzhou Metro App, enabling mixed use of QR codes and facial recognition ("Code-Face Mixed Use"). This allows for sub-second, sub-300-millisecond face recognition transit through gates, enabling a "One Face for the Whole Line" experience. This achieves truly seamless, stop-free rapid passage.

The launch of the nation's first metro VR large-scale free-roam area at Dongfeng Road Station on Zhengzhou Metro Line 2 further expands the functional boundaries of metro spaces. This project utilizes high-precision positioning systems and wireless wearable equipment, completely breaking the constraints of physical space and offering users an unprecedented immersive virtual reality experience. This innovation of a new "Experience-Consumption-Socialization" scenario signals the transformation of urban transport hubs into diversified, immersive experiential spaces.

Table 19-2 Comparative Analysis of Application Effectiveness: Digital Twin vs. AR/VR Technologies

Application Domain	**Technology Type**	**Implementation Outcome**	**Application Case**
Tunnel Inspection	Digital Twin + LiDAR	Achieved a 120-fold efficiency improvement compared to manual inspection; detection accuracy reaches 0.3 mm, enabling a shift from manual sampling to comprehensive, data-driven condition assessment.	Guangzhou Metro Tunnel Comprehensive Detection System
Equipment Maintenance	AR Remote Collaboration	Significantly reduces or eliminates the need for expert travel to site locations; enhances first-time fix rates and accelerates problem resolution by providing real-time, context-aware guidance to on-site technicians.	HiScene AR Smart Inspection System
Passenger Guidance	AR Real-Scene Navigation	Enables accessible route planning for passengers with	MTR (Shenzhen) Multifunctional

		mobility challenges (e.g., wheelchair users, those with strollers/luggage); provides intuitive, precise interaction via real-world overlay, improving wayfinding confidence and experience.	Smart Terminal
Staff Training	VR Immersive Experience	Allows personnel to realistically experience operational details and scenarios in a safe, controlled virtual environment; enhances training effectiveness, knowledge retention, and preparedness for real-world situations.	Guangzhou Metro Museum Interactive Exhibits
Predictive Maintenance	Digital Twin + Dynamic Simulation	Demonstrated potential to reduce operational and maintenance costs by up to 30% and extend the useful life of critical assets by enabling proactive, condition-based interventions before failures occur.	Rail Fastener Damage Prediction Model

19.5 Technology Integration and Future Development

19.5.1 Integration of Digital Twin and AR/VR Technologies

Guangzhou Metro has actively explored the integration of Digital Twin and AR/VR technologies. HiScene, as an ecosystem partner of SuiTeng Digital Technology, has created a new quality productive force through the deep integration of AR and AI, connecting rail transit with the digital world. This fusion has created entirely new operational maintenance and service models, bringing revolutionary changes to the metro system.

In the construction of the intelligent stations for Lines 10 and 12, PDT Science & Technology applied Digital Twin technology throughout the entire lifecycle of both lines, spanning construction, operation, and maintenance. This application ranged from commissioning control during the construction phase to simulation and decision support during the operational maintenance phase, effectively enhancing project quality and operational efficiency. This full-lifecycle application of Digital Twin provides a unified data environment and decision support for the planning, design, construction, and maintenance of metro facilities.

Looking ahead, the integration of Digital Twin and AR/VR technologies will become even tighter. The WebGIS-based Digital Twin platform provides an example for the industry's digital transformation through its 3D visualization and intelligent decision-making functions. The future will see further integration of Digital Twin and Artificial Intelligence (AI) technologies to enhance the autonomous diagnostic capabilities for track health status.

19.5.2 Intelligent Development Trends

With the continuous development of technology, the application of Digital Twin and AR/VR technologies in metro systems is expected to exhibit the following development trends:

- **Comprehensive Digitalization:** Utilizing high-precision sensors and 3D modeling technologies to achieve a digital representation of all metro elements, constructing a Digital Twin that is fully consistent with the physical metro.

- **Intelligent Decision-Making:** Leveraging artificial intelligence and big data analytics to achieve intelligent diagnosis and prediction of equipment status, supporting more scientific and precise maintenance decisions.

- **Human-Machine Collaboration:** Utilizing AR/VR technologies to achieve deep integration between humans and systems, improving work efficiency and service quality.

- **Platform Unification:** Utilizing a unified digital platform to integrate the entire process of planning and design, construction implementation, and operation and maintenance management, achieving data sharing and business synergy.

Guangzhou Metro has conducted valuable explorations in these areas, providing referential experience for the industry. In the future, as technologies mature and application scenarios expand, Digital Twin and AR/VR technologies will play an even more important role in metro systems.

19.6 Significance for International Vocational Education and Talent Cultivation

19.6.1 Updating the Vocational Education Curriculum

The development of Digital Twin and AR/VR technologies imposes new requirements on the vocational education curriculum for rail transit. There is a need to add new course modules such as Digital Twin Technology, AR/VR Applications, Big Data Analytics, and AI Algorithms to equip students with the skills to construct, manage, and apply Digital Twin systems.

Using Guangzhou Metro's Tunnel Comprehensive Detection System as an example, vocational education must cover the principles and applications of technologies like LiDAR scanning, 3D structured light measurement, and high-definition line-scan camera imaging, alongside methods for analyzing and processing the resulting inspection data. For instruction on the AR Smart Inspection System, the curriculum needs to include AR device operation, remote collaboration workflows, and fault identification and diagnosis.

19.6.2 Practical Training and Skills Development

Guangzhou Metro's practical experience with Digital Twin and AR/VR technologies provides rich case studies for practical teaching in rail transit vocational education. By establishing training platforms that replicate real operational maintenance environments, students can gain hands-on experience operating Digital Twin systems, analyzing operational data, and performing fault diagnosis and prediction. This cultivates practical operational abilities and problem-solving skills.

Specifically, practical teaching projects can be designed around areas such as:

- Analysis and processing of tunnel inspection data.

- Operation and maintenance of the AR Smart Inspection System.
- Construction and management of the Digital Twin platform.

Through this project-driven teaching approach, students can gain a deep understanding of the application scenarios and operational workflows of Digital Twin and AR/VR technologies, laying a solid foundation for future careers in intelligent rail transit maintenance.

19.6.3 International Certification and Standards Alignment

Guangzhou Metro's practical application of Digital Twin and AR/VR technologies provides a technical foundation and standard reference for the international certification of rail transit vocational education. By integrating Guangzhou Metro's technical standards for Digital Twin, application specifications for AR/VR, and personnel training requirements into vocational curricula, the cultivation of talent can be better aligned with international practices, thereby enhancing the global relevance of the vocational education provided.

Furthermore, when Guangzhou Metro's Digital Twin and AR/VR technologies are exported to international markets, they require corresponding support from technically skilled personnel. Through collaboration with international vocational education institutions, developing training courses and certification systems based on Guangzhou Metro's practical experience can help cultivate technical talents suited to the demands of the global market. This, in turn, promotes the international dissemination of China's rail transit technologies.

Chapter Summary

Guangzhou Metro's practices in applying Digital Twin and AR/VR technologies represent the current forefront of digitalization and intelligentization in urban rail transit. Through the construction and application of systems like the Tunnel Comprehensive Detection System, the intelligent operation and maintenance platform, the AR Smart Inspection System, and VR immersive experiences, Guangzhou Metro has achieved significant improvements in operational efficiency and noticeable enhancements in service quality. This provides valuable, transferable experience for the digital transformation of urban rail transit globally.

These technological innovations and practical experiences not only provide robust support for the high-quality development of Guangzhou Metro itself but also offer a valuable reference for the intelligent upgrade of urban rail transit systems worldwide. By establishing a systematic international vocational education system, these innovations and experiences can be applied and promoted more broadly, propelling the intelligent evolution of operation and maintenance technologies in the global urban rail transit sector. As technology continues to advance and application scenarios expand, the use of Digital Twin and AR/VR technologies in Guangzhou Metro will deepen further, offering even stronger technical guarantees for the safe, efficient, and green operation of urban rail transit.

Part V: Construction Innovation Technologies

Chapter 20: Technological Innovations in the Construction of Guangzhou Metro Line 12

20.1 Project Overview and Technological Innovation System

Guangzhou Metro Line 12 serves as a crucial "X-shaped" diagonal component within the framework of Guangzhou's urban rail transit network, functioning as a backbone line traversing the city center. The line spans 37.6 kilometers and comprises 25 stations, of which a remarkable 17 are interchange stations. This results in an interchange ratio of 68%, setting a new record for the highest proportion of interchange stations within the Guangzhou Metro system. The line connects several administrative districts, including Yuexiu, Liwan, Tianhe, and Baiyun, effectively linking diverse urban hubs such as major cultural and sports venues, central business districts, and key transportation interchanges. Line 12 achieved a significant milestone with the completion of its entire tunnel structure ("Tunnel Breakthrough") in May 2025. It subsequently commenced initial operation of its eastern and western sections (from Xunfenggang to Guangzhou Gymnasium, and from Ersha Island to University

Town South) on June 29, 2025.

Throughout its construction, Line 12 fostered the development of a comprehensive technological innovation system. This system encompasses four core domains:

1.Shield Tunneling Technology in Complex Geological Conditions

2.Construction Technology for Deep, Large-Scale, and Complex Stations

3.Precision Technology in Track Engineering

4.Smart Construction and Maintenance Technologies

These innovations have not only resolved a series of complex engineering challenges but have also accumulated valuable experience for metro construction within dense urban cores. Consequently, they hold significant pedagogical value for international vocational education in the field of civil and rail transit engineering.

Table 20-1 Basic Information of Guangzhou Metro Line 12

Item / Category	Parameters / Details	Remarks / Elaboration
Line Length	37.6 km	-
Number of Stations	25 Stations	-
Number of Interchange Stations	17 Stations	Interchange ratio as high as68%, significantly improving the accessibility and convenience of the entire metro network.
Deep & Large Station	Chigang Station (Depth: 41 m)	The deepest station along the entire line, reflecting the significant engineering challenges overcome in construction.
Tunnel Completion Date	May 2025	-
Commencement of Operation	June 29, 2025	Initial operation of the Eastern and Western sections. The specific timetable for the full line opening is to be confirmed.

20.2 Technological Innovations in Shield Tunneling under Complex Geological Conditions

20.2.1 Comprehensive Detection and Treatment Technologies for Karst Geology

The left-line tunnel of the Jingtai to Keziling section on Line 12 spans a total length of 1,033 meters. This segment is characterized by highly complex and diverse geological conditions, predominantly consisting of moderately to slightly weathered limestone and highly weathered shale. The encounter rate of karst caves within the tunnel alignment is notably high, reaching 21.5%, posing significant geotechnical challenges. Furthermore, the shield tunneling operation necessitated underpassing densely populated areas, including the Keziling building clusters, the Dajinzhong Road - Xinguangcong Road viaduct, and the Southern Baiyun Avenue viaduct. The presence of these multiple sensitive risk sources substantially increased the difficulty and complexity of the tunneling process.

To effectively address these intricate ground conditions, a dual-mode (slurry and earth pressure balance) shield tunneling machine was deployed for the excavation. The adaptability and robustness of the equipment in such challenging environments were enhanced through several key modifications: the cutterhead structure was reinforced, and the shield was equipped with advanced features such as a visualized soil chamber and probe drilling systems. For ground investigation, sophisticated techniques including micro-geophysical exploration combined with cross-hole CT imaging were employed to proactively detect and delineate karst cave development zones, enabling preemptive treatment. An innovative approach involving down-the-hole hammer drills to create inclined boreholes was adopted specifically for treating cavities beneath existing buildings, maximizing the reinforcement of the building foundations. For sections traversing beneath bridge piles, a pile foundation underpinning technique was successfully implemented, ensuring the structural safety and stability of the bridges throughout the tunneling process.

20.2.2 Precision Control and Safety Assurance in Shield Tunneling

Throughout the shield tunneling construction of Line 12, a comprehensive quality control system was established and rigorously implemented. This system integrated real-time monitoring with dynamic adjustment mechanisms to ensure stringent control over the tunneling axis and alignment accuracy. Recognizing the varying ground conditions along the route, differentiated sets of tunneling parameters were meticulously formulated and applied for different geological zones. This strategy enabled precise control over the shield attitude (yaw, pitch, roll) and minimized deviations.

Particularly in critical sections involving underpassing of major structures, an automated monitoring system was deployed. This system provided real-time, high-frequency (minute-level updates) data on the deformation behavior of the existing structures. The capability for near-instantaneous data acquisition and analysis was crucial, allowing for the timely detection and mitigation of any anomalies during the shield passage. The successful application of this integrated monitoring and control approach across multiple underpassing sections on Line 12 provides a valuable and reliable technical reference for similar future projects facing analogous challenges.

Table 20-2 Key Technological Innovation Points for Addressing Complex Geological Conditions in Line 12

Technology Type	Technical Characteristics	Application Section	Implementation Effects
Dual-Mode Shield Tunneling Technology	Enables flexible switching between Slurry Mode and Earth Pressure Balance (EPB) mode, providing high adaptability to significant variations in ground conditions and composite strata.	Jingtai to Keziling Section	Enhanced machine versatility and operational continuity, significantly reducing tunneling risks associated with heterogeneous geology and lowering the potential for machine entrapment or face instability.

Comprehensive Geological Detection	Integrates Micro-geophysical Prospecting with Cross-hole CT (Computerized Tomography) imaging to achieve high-precision, three-dimensional detection and characterization of karst caves and cavities.	Karst Development Zones along the entire line	Enabled safe and efficient tunnel advancement despite a high karst cavity encounter rate of 21.5%, allowing for proactive ground treatment and minimizing unforeseen hazards.
Inclined Grouting Technique	Utilizes Down-The-Hole (DTH) hammer drills to create precisely angled (inclined) boreholes for targeted grouting, effectively reinforcing building foundations directly beneath structural footprints.	Sections Underpassing Existing Buildings	Successfully minimized ground settlement and differential deformation beneath structures, ensuring the structural integrity and safety of overlying buildings during tunneling operations.
Pile Foundation Underpinning Technique	Involves constructing temporary support structures followed by the systematic transfer of loads from existing bridge piles to new, deeper foundations, ensuring seamless transition and permanent structural stability.	Sections Underpassing Bridges	Guaranteed the absolute safety and uninterrupted service of bridge structures above, preventing any adverse effects on traffic and infrastructure due to tunneling-induced subsidence.
Automated Real-time Monitoring	Employs a network of sensors with automated data acquisition and transmission, providing	Critical Underpassing Sections	Facilitated the prompt detection of minute structural deformations,

	continuous, real-time monitoring data updated at minute-level intervals for immediate analysis and response.	enabling immediate intervention and adjustment of tunneling parameters to mitigate risks and ensure construction safety.

20.3 Innovations in the Construction Technology of Deep and Complex Stations

20.3.1 Half-Tunnel Column Method with Close-Proximity Underpassing Technology at Chigang Station

Chigang Station, the deepest station on Guangzhou Metro Line 12 at 41 meters, features an excavation area of approximately 14,000 square meters and serves as a three-line interchange hub. The station's surrounding environment is extremely complex. Due to phased land acquisition and demolition, as well as the need to overpass and underpass existing lines, the entire station and its ancillary excavations were segmented into eight irregularly shaped foundation pits of varying sizes.

The station employed the nationally pioneering "Half-Tunnel Column Method with Close-Proximity Underpassing Technology for Existing Lines." This technique integrates concepts from both cover-cutting and underground excavation methods. It utilizes six newly constructed support pier columns for vertical support and the existing tunnel structure as a protective "cover." This system rigidly locks the spatial position of the existing tunnel, allowing underground excavation to proceed safely beneath this protective cover. Faced with asymmetric excavation on both sides of the existing line and the consequent eccentric horizontal loading on the existing structure, the project team optimized the support system and managed load transfer rationally. This achieved millimeter-level deformation control of the existing operating line.

During the construction of Chigang Station, WSS grouting employed a "Dual-Fluid Variable Grouting Pump + Intelligent Grouting Monitoring and Protection System." This effectively reduced settlement of the existing structure during excavation. To ensure tight contact between the secondary lining roof slab of the underground excavation and the base slab of the existing line, the team used self-compacting, ultra-retarding, micro-expansive concrete combined with a bridge bearing grouting process. This approach effectively mitigated long-term settlement of the existing structure and enhanced its safety throughout its entire lifecycle.

20.3.2 BIM Technology and Intelligent Fire Protection System at Chigang Tower Station

As one of the characteristic stations on the line, Chigang Tower Station boasts a complex design with thousands of square meters of double-curved aluminum panel cladding, requiring high precision in material fabrication. To ensure construction efficiency and quality, Guangzhou Metro, in collaboration with the constructor, utilized BIM technology to optimize the 3D layout of the decorative surfaces, achieving "zero-error" pre-assembly. Furthermore, resources from multiple manufacturers were coordinated, adopting a "multi-factory simultaneous production scheduling + resident factory supervision" model to strictly control fabrication quality. During the installation in public areas, technical personnel worked in shifts to ensure the first-time successful installation of the double-curved aluminum panels, resulting in a smooth visual effect.

Addressing the challenge of commissioning fire protection equipment in the station's 14-meter-high large space, the constructors implemented an intelligent fire protection system comprising "Dual-Technology Detectors + Fire Water Monitors." Utilizing composite infrared and image detection technology, this system enables precise fire source localization and rapid response. Confronted with the complex commissioning environment and difficult signal coverage in such a large space, Guangzhou Metro collaborated with equipment suppliers on a specialized solution. They adopted a "simulated working conditions + layered commissioning" strategy, ensuring the efficient completion of equipment debugging.

20.3.3 Prefabricated Construction and Multi-Disciplinary Integrated Commissioning

Line 12 extensively employed prefabricated construction techniques in station building. Modular installation was used for equipment in key areas, such as fire pump rooms, reducing on-site workload by approximately 40% and significantly enhancing construction efficiency. Facing the pressure of numerous commissioning points and complex interfaces, the team also established a "checklist-based" commissioning ledger. They conducted "Multi-Disciplinary Integrated Commissioning" in collaboration with supervision and operation units, ensuring a smooth handover.

Through a 3D simulation system, various adverse working conditions were analyzed numerically, followed by multiple rounds of scheme validation and risk assessment. This process ultimately led to the successful breakthrough of nearly 10,000 meters of shield tunnels In the construction of the 142-meter underground excavation tunnel between Chigang Tower and Chigang stations, the project team innovatively adopted a hybrid method combining "Mechanical excavation on the upper bench + Controlled blasting on the lower bench," effectively controlling rock mass disturbance. The project established a comprehensive "Dust - Noise - Emissions" prevention and control system. A self-developed assembled acoustic enclosure, designed with soundproofing cotton and diversion vents, resolved noise issues from dust removal equipment. This provides a exemplary model for underground construction in environmentally sensitive urban areas.

20.4 Precision Track-Laying Technologies in Rail Engineering

20.4.1 Green and Efficient Track-Laying Technology

The track-running area from Xunfenggang to Yunxi Park (exclusive) on Line 12 spans approximately 12.9 kilometers and incorporates five distinct types of track bed structures. This presented significant challenges, including frequent transitions between track bed types, considerable construction difficulties, and complex organizational coordination. Since track-laying commenced in August 2023, Guangzhou Metro and the participating construction units rationally allocated production resources, establishing a total of six track-laying bases. During peak construction periods, work was conducted simultaneously across 11 working faces, involving over 500 personnel.

The track-laying operation deployed more than 50 pieces of equipment powered by new energy sources, including track-laying machines, rail vehicles, and rail welders. These equipment choices featured high energy utilization efficiency and a close match between power and load, effectively reducing overall energy consumption and enhancing the project's green construction credentials. As the track-laying bases were situated near residential areas, noise-reducing panels were installed in equipment processing zones to effectively mitigate construction noise. Furthermore, the constructors innovatively implemented a track zone management system and a self-developed cloud-based rail management platform. These digital tools enabled quantitative management of various energy consumptions, rational control of material usage, improved construction efficiency, and reduced safety risks.

20.4.2 High-Speed Railway Standard Precision Track Adjustment Technology

During the track-laying process, the metro construction team conducted technical research into the precise adjustment of long rails. They utilized a Grade 0 track inspection car to measure various track geometry parameters, including alignment coordinates, elevation, gauge, and cross-level. Based on the collected data, the required track adjustment values were analyzed and calculated to formulate an optimal fine-tuning plan. Simultaneously, the team strictly implemented standardized operational procedures during on-site execution and conducted comprehensive inspections of the fine-tuning work quality. This rigorous approach rapidly and accurately enhanced the track's high-level smoothness (geometric quality). During the hot-running test, this achievement was vividly demonstrated by the ability to stand a coin upright on its edge on a straight section without it falling over while the train was in motion—a result approaching the smoothness standards of high-speed railways. This success establishes a solid foundation for the safe and comfortable operation of the metro line.

20.5 Innovation in Intelligent Construction and Operation & Maintenance Technologies

20.5.1 Building Information Modeling (BIM) Application Across the Entire Lifecycle

Line 12 comprehensively implemented 3D Building Information Modeling (BIM) technology throughout the project. This enabled the advanced simulation of integrated utility layouts, optimizing and resolving clash detection prior to physical construction. A strict "Prototype First, Layered Inspection" regime was enforced, ensuring millimeter-level precision in equipment installation and achieving the goals of "fewer clashes and less rework." For instance, at University City North Station, where construction involved interface challenges with the existing Line 4 for interchange modifications and co-construction with commercial developments, Guangzhou Metro collaborated with design, supervision, and other stakeholders. They optimized the construction scheme through strategies like "phased zonal construction" and "staggered peak-hour work schedules," thereby rationalizing site arrangement and sequencing.

To guarantee high-quality handover, the construction and operations departments established a "Dual-Early" linkage mechanism, adhering to the principle of "Early Optimization, Early Inspection, Early Rectification." Multiple specialized coordination meetings were conducted. Measures such as the early involvement in equipment commissioning and acceptance significantly strengthened on-site supervision and follow-up. This ensured the efficient completion, according to schedule, of tasks including the installation, commissioning, and acceptance of signaling systems and low-voltage electrical systems within the handover areas.

20.5.2 Intelligent Stations and Fully Automatic Operation (FAO) System

Stations on Line 12 are equipped with a newly upgraded suite of intelligent travel products independently developed by Patech (Guangzhou Patech Technology Co., Ltd.). These products provide passengers with smart services throughout their journey, including intelligent inquiry and seamless, frictionless fare payment. The payment zones of the smart fare gates integrate CRcode, NFC, and Face ID technologies, while also adopting new paddle gate units, creating more convenient and secure passenger passageways.

The new-generation upgraded smart customer service centers integrate technologies such as AI voice recognition and biometrics. Passengers can independently complete various self-service functions, including ticket transaction processing and information inquiries. Passengers can register their facial data at these centers to activate the facial recognition fare gate feature. By binding with the Guangzhou Metro APP ride code, the system enables "mixed use of code and face," allowing passengers to experience millisecond-level facial recognition for gate passage. This enables a "one-face pass" across the entire line, with a facial recognition latency of less than 0.3 seconds, truly achieving a seamless, non-stop rapid transit experience.

Line 12's Guangzhou Baiyun Station facilitates seamless interchange with the national railway station of the same name. Passengers can transfer between them within the station in approximately five minutes, benefiting from a one-way security check exemption when transferring from the national railway to the metro. The Automatic Fare Collection (AFC) system, supplied by Patech, ensures the rationality and accuracy of the ticketing system through technical means, while passengers enjoy this convenient transfer service. Specifically, between the high-speed rail station and the metro station at Guangzhou Baiyun Station, new transfer gates provided by Patech eliminate the need for a secondary security check, allowing travelers to transfer effortlessly between high-speed rail and metro services.

20.5.3 Intelligent Operation & Maintenance and Whole-Lifecycle Management

Patech's "Jia Yi Wei" Platform Door Intelligent Operation and Maintenance system has been piloted on Line 12. Leveraging big data, the Internet of Things (IoT), and cloud computing technologies, it provides more accurate and timely real-time status monitoring, fault prediction, and alarm functions for equipment facilities. This enables a precise shift from scheduled maintenance to "Condition-Based Maintenance" (CBM) while ensuring operational safety. The system enhances management efficiency and reduces costs over the entire equipment lifecycle, improving overall O&M efficiency and effectively lowering O&M expenses.

Both lines of the initial operating section employ the highest Grade of Automation (GoA4) Fully Automatic Operation (FAO) system. This achieves unattended operation throughout the entire process, including train wake-up, self-testing, running, and dormancy, representing a historic breakthrough for Guangzhou rail transit. To ensure the safe operation of automated trains, Patech implemented a comprehensive upgrade of the platform door system. By assigning independent identification numbers, each platform and every platform door is given a unique "ID." This allows for rapid and accurate localization of specific doors, whether for routine inspections, fault troubleshooting, or remote real-time monitoring.

Table 20-3 Innovative Applications of Smart Technologies on Line 12

Technology Domain	**Innovative Technology**	**Functional Characteristics**	**Application Outcomes**
Smart Station	One-Click Station Startup/Shutdown System	Integrates and coordinates over 20 subsystems (e.g., lighting, environmental control, fare gates, passenger information systems, security) across different equipment platforms, enabling centralized, automated control of station operational states.	Improves daily operational efficiency by approximately 30% by drastically reducing manual procedures and time required for station opening and closing, while minimizing human error.
Intelligent Service	Millisecond-Level Facial Recognition	Enables high-speed identity verification with a recognition latency of <0.3 seconds. Supports flexible "Mixed Code & Face" authentication, allowing passengers to use either payment method interchangeably.	Facilitates seamless, frictionless access through fare gates, significantly enhancing passenger convenience, travel flow, and overall journey experience by eliminating physical interactions and stops.
Intelligent Maintenance	Patech 'Jia Yi Wei' Platform Door System	Employs IoT sensors and data analytics to provide real-time status monitoring, performance analysis, and proactive fault early warning for platform screen doors, enabling data-driven maintenance	Implements a shift from preventive to Condition-Based Maintenance (CBM), effectively reducing unplanned downtime,

		decisions.	extending equipment lifespan, and lowering long-term operation and maintenance costs.
Fully Automatic Operation	GoA4-grade FAO System	Achieves the highest Grade of Automation (GoA4), enabling fully unattended operation of trains throughout all phases, including wake-up, self-testing, revenue service, turnaround, and shutdown.	Enhances operational safety and service reliability by eliminating human operational error, enables optimized traffic management, and allows for more flexible and responsive train scheduling.
Passenger Flow Management	AI-driven Multi-modal Passenger Flow Prediction	Utilizes Artificial Intelligence (AI) and data fusion techniques to analyze and forecast real-time and future passenger flow patterns, supporting dynamic, data-informed decision-making for crowd management measures like station throttling.	Enables precise and proactive passenger flow management, effectively alleviating congestion hotspots, optimizing passenger distribution, and improving station safety and comfort levels, especially during peak hours.

20.6 The Significance of Line 12's Technological Innovations for International Vocational Education

The technological innovations implemented during the construction of Guangzhou Metro Line 12 provide a wealth of teaching resources and case study references for international vocational education in the urban rail transit sector. These practical, project-based technological advancements not only showcase China's latest achievements in urban rail transit construction but also offer invaluable content for international vocational training curricula.

20.6.1 Vocational Curriculum Development and Teaching Resource Construction

The technological innovations from Line 12 serve as rich source material for updating vocational education curriculum systems for rail transit. Courses on tunneling in complex geology can incorporate the comprehensive karst detection and treatment technologies employed on Line 12. In courses covering deep and large station construction, the Half-Tunnel Column Method with Close-Proximity Underpassing Technology used at Chigang Station presents an exceptional teaching case. For courses on intelligent metro systems, Line 12's Fully Automatic Operation (FAO) system and its intelligent maintenance technologies represent the industry's cutting-edge development trends.

Building upon the practical experience gained from Line 12, virtual simulation teaching resources can be developed. These would create immersive virtual training environments for key processes such as tunnel construction, station building, and equipment commissioning. Utilizing VR/AR technologies, students can gain firsthand experience with critical construction techniques under complex working conditions, thereby deepening their understanding of theoretical knowledge. Furthermore, the actual BIM models created for Line 12 can be directly converted into teaching models, providing students with clear demonstrations of whole-lifecycle management in metro engineering projects.

20.6.2 International Talent Cultivation and Standards Export

The technological innovations of Line 12 provide a reference for technical standards in cultivating international talent for the rail transit industry. By integrating elements such as Line 12's technical standards, operational procedures, and safety management protocols into vocational education programs, it is possible to cultivate technical skilled personnel who meet the demands of the international market. Particularly, the innovative practices in intelligent metro systems implemented on Line 12 supply forward-looking technological content for international vocational education.

As Chinese rail transit technology continues to expand globally, the construction experience of Line 12 will become a vital resource for international vocational education collaboration. Through initiatives such as training programs in China and establishing overseas educational partnerships, the technological achievements of Line 12 can be promoted on the international stage. This approach helps cultivate localized talent who understand Chinese standards and master Chinese technologies, thereby facilitating the international promotion of Chinese rail transit expertise.

Chapter Summary

The technological innovations in the construction of Guangzhou Metro Line 12, spanning the four major fields of shield tunneling in complex geology, deep and complex station construction, precision track-laying engineering, and intelligent construction and operation & maintenance, collectively form a comprehensive technological system for metro construction in urban core areas. These innovations have not only successfully resolved the technical challenges encountered during the Line 12 project, ensuring its safe and high-quality completion, but also provide transferable experience for urban rail transit construction under similar conditions worldwide.

The technological achievements of Line 12 enrich the technical framework of urban rail transit construction and offer significant reference points for the industry's technological progress. Simultaneously, these practice-based innovative technologies supply authentic, cutting-edge teaching cases and technical standards for international vocational education within the rail transit sector, holding substantial value for cultivating globally competent technical and skilled talent. With the commencement of operations on Line 12 and the broader application and promotion of these innovative technologies, they are poised to contribute more "Guangzhou Wisdom" and "Chinese Solutions" to global urban rail transit development.

Chapter 21: Guangzhou Metro Line 18: Innovative Breakthroughs and the Future of Travel

21.1 Line Overview and Strategic Positioning

Guangzhou Metro Line 18, serving as a strategic backbone for the rail transit network within the Guangdong-Hong Kong-Macao Greater Bay Area (GBA), stands as China's first fully underground regional express line designed for an operational speed of 160 km/h. This line not only sets a new record for metro operational speed in the country but also represents comprehensive breakthroughs in construction standards, technological equipment, and operational models. It embodies the highest current level of achievement in China's urban rail transit development, signaling a significant leap forward for the industry.

21.1.1 Line Planning and Technical Parameters

Line 18 adopts a composite functional positioning, integrating the roles of a core urban metro service within central Guangzhou and a high-speed regional express link connecting to outlying districts. It strategically connects Guangzhou's city center with the Nansha New District, a key area for future development. Critically, the line's planning incorporates provisions for future extension to Zhongshan and Zhuhai, enhancing its long-term role in regional integration within the GBA. Its design successfully merges the advantages of both traditional metro systems and intercity railways. This unique synthesis achieves an optimal combination of "High-Frequency, Metro-like Operation" and "Rapid Regional Connectivity," offering passengers both the convenience of frequent urban services and the speed necessary for longer-distance travel across the metropolitan region.

Table 21-1 Fundamental Technical Parameters of Guangzhou Metro Line 18

Item	Parameter / Specification	Remarks / Elaboration
Design Speed	160 km/h	Represents the highest design and operational speed record for a metro system in China, positioning it as a true regional express rail service.
Operational Mode	Combination of Express & Local Services	Implements a skip-stop pattern where express trains overtake local ones, significantly enhancing line capacity, optimizing travel times for long-distance passengers, and improving overall operational efficiency.
Grade of Automation (GoA)	GoA4 (Fully Automatic Operation)	The highest Grade of Automation, enabling unattended operation throughout all phases, including train wake-up, running, door handling, and shutdown, thereby enhancing safety and operational precision.
Catenary Type	Rigid Overhead Catenary System	Specifically selected for its superior stability and current collection quality at high speeds compared to flexible systems, ensuring reliable power transmission and pantograph contact at 160 km/h.
Through Operation	Interconnection with Guanghua, Fangbai, and Nanzhu (Zhongshan) Intercity Railways	Serves as a pioneering demonstration project for the integrated 'Four-Networks Fusion' strategy, seamlessly connecting metro, suburban railway, and intercity rail networks to create a unified regional transit system.

The route planning of Line 18 demonstrates significant foresight. Its northern extension (Guanghua Intercity Railway) will connect Guangzhou East Railway Station to Tian'gui in Huadu District, while the southern extension (Nanzhu-Zhong Intercity Railway) will integrate Zhongshan and Zhuhai, ultimately forming a strategic north-south transportation artery traversing the western flank of the Greater Bay Area.

As of June 2025, construction progress on the Guanghua Intercity section (the northern extension of Line 18) has been substantial, with civil engineering works 48% completed. All seven tunnel sections along this segment have now fully entered the shield tunneling phase, demonstrating accelerated project advancement. Notably, 23 shield tunneling machines are currently operating concurrently, highlighting the immense scale and coordinated intensity of the construction efforts.

21.1.2 Strategic Significance in the Greater Bay Area Context

Line 18 serves as a critical pillar for realizing the "One-Hour Commuting Circle" within the Guangdong-Hong Kong-Macao Greater Bay Area (GBA). Upon its full completion, travel times will be dramatically reduced: the journey from Zhujiang New Town to Wanqingsha in Nansha will take approximately 28 minutes, while the connection from Guangzhou East Railway Station to Nansha will be achieved in around 30 minutes. This effectively "compresses" the spatiotemporal distance between the city's core and its key sub-centers, fostering unprecedented accessibility.

More significantly, Line 18 achieves a genuine level of "Four-Networks Integration" through its through-operation design. Acting as a strategic backbone, it seamlessly interconnects the Nanzhu (Zhong) Intercity Railway, the Guanghua Intercity Railway, the Fangbai Intercity Railway, and Guangzhou Metro Line 22. This integrated network functionally links major urban hubs across the western GBA, including Zhuhai, Zhongshan, Nansha, Panyu, Baiyun, and Huadu districts of Guangzhou. Consequently, Line 18 acts as a powerful catalyst, accelerating the GBA's transition from a collection of distinct cities towards an integrated, functionally unified metropolitan region characterized by "homogenized living."

21.2 Core Technological Breakthroughs and Innovative Applications

21.2.1 Variable-Diameter Shield Tunneling Machine: Station-Tunnel Integrated Construction

The "Variable-Diameter No. 1," the world's first in-situ variable-diameter shield tunneling machine developed under the leadership of Guangzhou Metro Group, represents a major technological breakthrough in the construction of Line 18. This equipment successfully addresses the global challenge of integrated station-tunnel construction for rail transit in densely built urban cores.

Technical Principles and Innovative Features:

(1) In-Situ Diameter Variation: The shield machine can modify its excavation diameter directly at the predetermined location during the tunneling process, eliminating the need for pre-excavated launch shafts traditionally required for diameter changes. This overcomes significant limitations of conventional variable-diameter equipment.

(2) Continuous Stepless Adjustment: The cutterhead diameter can be adjusted steplessly within a range of 8.83 meters to 12.45 meters, capable of both increasing ("small-to-large") and decreasing ("large-to-small"). During this process, the excavation area expands by approximately 98%, readily accommodating the large spatial requirements of station tunnels.

(3) System Integration: Through proprietary technologies such as synchronous retraction/extension of multiple cutterbeams and heavy-duty variable-diameter segment assembly in confined spaces, the machine achieves intelligent, coordinated control over cutterhead diameter variation, shield body diameter adjustment, spatial protection, and power matching.

(4) Application Value: The "Variable-Diameter No. 1" was first deployed in the Guangzhou Railway Station project of the Fangbai Intercity Line. It enabled a novel construction mode described as "One Station, Two Tunnels, Formed Continuously." This innovative method significantly reduces the impact on surface traffic and environmental disturbance associated with open-cut excavation, offering a new

"Chinese Solution" for rail transit construction in urban core areas.

21.2.2 Rigid Overhead Catenary System: Ensuring High-Speed and Stable Current Collection

As China's first fully underground regional express line operating at 160 km/h to utilize a rigid overhead catenary system, Line 18 represents a significant innovation in its traction power supply system.

Technical Challenges and Solutions: High-speed metro catenary systems must resolve the stability issues of traditional flexible suspensions during current collection at very high speeds. For Line 18, a new type of cantilever measurement device was developed. By enhancing measurement accuracy and efficiency, this device ensures the stability and reliability of the rigid catenary system under high-speed operational conditions.

Key Technological Innovations:

(1) **High-Precision Measurement:** Addressing the problems of significant errors and low efficiency in existing cantilever measurement methods, a new measurement methodology based on the novel cantilever measurement device was introduced. This innovation substantially improves the efficiency and accuracy of measuring the deflection and geometry of the rigid overhead catenary installation.

(2) **Reliable Current Collection:** The rigid catenary system provides a continuous and stable power supply to trains operating at high speeds, ensuring safe and reliable operation at the target speed of 160 km/h.

21.2.3 Fully Automatic Operation System: Highest Grade of Automation

Line 18 employs the GoA4 (Grade of Automation 4) Fully Automatic Operation technology (also referred to as FAM mode), representing the highest grade of automation in urban rail transit.

System Functions:

(1) **Automatic Operation:** The system supports fully unattended operation throughout the entire process, including train wake-up, self-testing, movement in/out of depots, departure from terminals, station stopping, door opening/closing, and shutdown (sleep mode).

(2) **Intelligent Dispatching:** Through seamless integration between the signaling system and the control center, the system enables precise control and flexible adjustment of train operations and scheduling.

(3) **Operational Advantages:** The FAOS not only reduces labor costs but also, through its intelligent operational control, enhances operational efficiency and service reliability. Documented video evidence from the line shows trains operating smoothly in FAM mode, with platform stopping accuracy achieving centimeter-level precision.

Table 21-2 Function Grades of the Fully Automatic Operation System for Line 18

Grade of Automation (GoA)	Control Mode	Technical Characteristics	Application Scenario
GoA4 (Fully Automatic Operation)	Unattended Train Operation (UTO)	The system manages all functions, including starting, stopping, door operation, obstacle detection, and handling emergencies, with no staff required on board.	Primary mode for mainline revenue service and all operations within depots and stabling yards. Represents the highest level of integration and operational automation.
GoA3 (Attended Fully Automatic Operation / DTO)	Driverless Train Operation	The system performs all driving and door operations automatically. However, a train attendant is present primarily for monitoring, customer service, and to intervene in exceptional circumstances.	Serves as a backup mode for specific scenarios or special working conditions, providing an additional layer of human oversight where deemed necessary.
GoA2 (Semi-Automatic Operation / STO)	Semi-Automated Train Operation	The system controls automatic acceleration, cruising, and braking. The	Typically used as a degraded or fallback mode of operation if

	train operator remains responsible for initiating departure, controlling door opening/closing, monitoring the platform, and handling any emergencies or anomalies.	failures occur in higher-grade systems, ensuring a basic level of service can be maintained.

21.3 Intelligent Maintenance and Efficient Operation Systems

21.3.1 SuiTeng OS Intelligent Rail Transit System

Line 18 comprehensively implements the "SuiTeng OS," a pioneering intelligent operating system recognized as the innovative "brain" of China's rail transit sector. This system was notably exhibited at the China Pavilion during the 2025 Osaka Expo in Japan, demonstrating Guangzhou Metro's technological capabilities on a global stage.

System Architecture and Functionality: The SuiTeng OS centers around an "Urban Transport Brain" concept. Utilizing Thing Model mapping and Digital Twin technologies, it integrates real-time data from 8 distinct train operation scenarios and 12 categories of equipment data. This integration facilitates dynamic simulation and supports intelligent decision-making processes. Its core functionalities include:

(1) Frictionless Fare Gate Transit: Supports multiple payment methods, including facial recognition and QR codes, enabling swift and seamless station entry and exit.

(2) Intelligent Passenger Flow Management: Automatically generates optimized crowd management strategies based on real-time passenger flow data, significantly enhancing the efficiency of passenger organization, especially during peak hours.

(3) One-Click Station Startup/Shutdown: Through deep system integration, consolidates numerous manual procedures traditionally performed sequentially by station staff into a single automated command, drastically streamlining daily operations.

(4) Video-Based Patrols and Inspections: Employs video analytics technology to automate security and facility inspections, replacing manual rounds and improving the efficiency and coverage of safety monitoring.

(5) Application Effectiveness: The deployment of SuiTeng OS has yielded significant operational benefits. Data indicates that its intelligent maintenance applications for electromechanical equipment have reduced the required maintenance personnel ratio by 16%. Manual station inspection rounds have been shortened from approximately 1 hour to just 10 minutes. In emergency response scenarios, the system demonstrated remarkable agility, completing tasks such as faulty line switching and generating passenger flow guidance schemes within a mere 15 seconds.

21.3.2 Intelligent Maintenance and Energy-Saving Technologies

Line 18 incorporates numerous intelligent maintenance and energy-saving technologies into its vehicle and system design, achieving optimized whole-lifecycle costs.

(1) **Permanent Magnet Traction System:** Utilizes Permanent Magnet Synchronous Motor (PMSM) traction systems. Compared to traditional asynchronous motors, this technology reduces energy consumption by approximately 15-20% while simultaneously lowering maintenance requirements due to its simpler and more robust design.

(2) **Inverter Air Conditioning Technology:**Automatically adjusts cooling capacity based on real-time passenger load and ambient temperature within the carriages. This enables precise environmental control and contributes significantly to energy conservation and consumption reduction.

(3) **Intelligent Maintenance System:**Continuously monitors the status of critical train systems. Leveraging big data analytics, it predicts potential equipment failures, facilitating a strategic shift from scheduled "Preventive Maintenance" to more efficient and cost-effective "Predictive Maintenance."

21.4 Future Travel Experience and Urban Integration

21.4.1 Cross-Modal Seamless Travel

Through its "Four-Networks Integration" technology, Line 18 achieves seamless connectivity between metros, regional railways, intercity railways, and mainline (national) railways. Passengers disembarking from a high-speed rail service can transfer seamlessly to a regional rail service. The implementation of a unified "One-Code Pass" system across all public transit modes allows passengers to genuinely experience a closed-loop journey where "Travel is Life."

This cross-modal seamless travel breaks down the traditional barriers between different transportation systems. Passengers can transfer conveniently without the need for multiple ticket purchases or repeated security checks, vastly improving travel efficiency. During demonstrations at the Osaka Expo, visitors interacting with this feature via touchscreens reportedly expressed admiration with comments like, "So advanced! This is the future of life in China!"

21.4.2 Human-Centric Service Design

Line 18 embodies a human-centric philosophy in its service design, introducing several innovative service features:

- **Express & Local Service Combination:**Operates a mix of express and local trains to cater to diverse passenger needs. Express trains serve major hubs for faster long-distance travel, while local trains provide comprehensive service to all stations along the line.

- **Business/Dedicated Carriages:** Plans are underway for the northern extension to introduce carriages tailored for business travelers. Research and design are also focused on integrating city terminal facilities (City Air Terminal) and baggage check-in systems, offering convenient pre-flight services for air travelers directly from metro stations.

- **Smart Restroom Guidance:** Utilizes a smart system that displays real-time occupancy status of restroom facilities, guiding passengers to available stalls quickly and reducing queueing times.

21.5 Implications for International Vocational Education

The technological innovations and practical implementation experience of

Guangzhou Metro Line 18 provide a rich repository of teaching resources and case study material for international vocational education within the rail transit sector.

21.5.1 Curriculum System Updates

Building upon the technical practices demonstrated by Line 18, vocational education curriculum systems should be updated to incorporate the following new content areas:

(1) **Principles and Operation of Variable-Diameter Shield Tunneling:** This module should cover the fundamental principles of in-situ shield diameter variation, the associated control systems, and the specific construction processes and sequences involved in its application.

(2) **High-Speed Metro Catenary Technology:** Courses need to include the design principles, installation methodologies, precise measurement techniques, and key maintenance requirements specific to rigid overhead catenary systems used in high-speed (160 km/h) underground environments.

(3) **Operation and Maintenance of Fully Automatic Operation Systems:** Training must encompass the standard operational procedures, fault diagnosis and response protocols, and emergency management strategies specific to GoA4-grade unattended train operation systems.

(4) **Intelligent Rail Transit Platform Management:** This area should address the system architecture, core functionalities, data integration mechanisms, and daily management practices of sophisticated operating systems like SuiTeng OS.

21.5.2 Training Platform Development

The technological advancements of Line 18 offer clear direction for the development of modern vocational training platforms:

(1) **Virtual Simulation Training:** Leveraging the detailed BIM models and Digital Twin systems developed for Line 18, dedicated virtual simulation training modules can be created. These would allow trainees to practice complex tasks such as variable-diameter shield tunneling operations and the dispatching and control of FAOS in a risk-free, virtual environment.

(2) **Real-Scenario Practical Training Platforms:** Constructing physical

training platforms that simulate real-world conditions is crucial. These should include FAOS simulator cabins for "driverless" operation training and setups for the precise measurement, adjustment, and debugging of rigid catenary systems, thereby cultivating hands-on operational skills.

21.5.3 International Certification System

The technical standards established and validated by Line 18 can serve as a valuable reference for developing international vocational certifications:

(1) **Export of Technical Standards:** Core technical standards developed for Line 18—encompassing variable-diameter shield tunneling, rigid overhead catenary systems, and FAOS—should be systematically integrated into the frameworks of international vocational qualification systems.

(2) **Mutual Recognition of Skill Certifications:** There is significant value in promoting the international mutual recognition of skill certifications that are based on these proven Chinese technical standards. This initiative would help cultivate a global workforce of technical skilled personnel who are familiar with and competent in applying these advanced technologies, thereby meeting the evolving demands of the international market.

Table 22-3 Correspondence Between Technological Innovations of Line 18 and Their Applications in Vocational Education

Technological Innovation	Corresponding Vocational Course	Practical Training Project	Potential Skill Certification
Variable-Diameter Shield Tunneling	Station-Tunnel Integrated Construction Technology: Covers the principles, equipment (like Variable-Diameter No. 1), control systems, and construction sequencing for this innovative method.	Virtual Simulation of Shield Diameter Variation: A hands-on module using BIM/Digital Twin models to practice the complex process of in-situ diameter adjustment in a risk-free virtual environment.	Intelligent Underground Construction Engineer: Certification validating expertise in advanced, automated, and integrated tunneling construction techniques.

Rigid Overhead Catenary System	High-Speed Catenary Technology & Measurement: Focuses on the design, installation, precise measurement techniques, and maintenance protocols specific to rigid catenaries for 160 km/h operations.	Catenary Precision Measurement Training: Practical sessions using specialized devices (e.g., the new cantilever measurer) to achieve the stringent accuracy required for high-speed current collection.	High-Speed Railway Catenary Technician: Certification for skills in installing, maintaining, and troubleshooting rigid overhead catenary systems.
Fully Automatic Operation (GoA4)	GOA4 System Principles & Maintenance: Details the architecture, operational protocols, failure mode analysis, and emergency management for unattended train operation systems.	FAM Mode Simulator Training: Training using high-fidelity simulators to experience and manage normal and degraded operations in a fully automated environment.	Fully Automatic Operation System Maintenance Specialist: Certification demonstrating competency in operating, monitoring, and maintaining GoA4-level FAOS.
SuiTeng OS Platform	Intelligent Rail Transit System Management: Explores the architecture of "urban transport brains," data integration (via Thing Models/Digital Twins), and management of core smart station functions.	Intelligent Dispatch Simulation Training: Practical exercises using a simulated SuiTeng OS environment to manage integrated operations, passenger flow, and respond to incidents.	Intelligent Rail Transit Systems Engineer: Certification for professionals capable of managing and optimizing complex, AI-enhanced rail transit operating platforms.

Chapter Summary

Guangzhou Metro Line 18 stands as a comprehensive embodiment of technological innovation within China's urban rail transit sector, achieving a series of breakthroughs in construction techniques, equipment manufacturing, operational management, and service models. The variable-diameter shield tunneling technology successfully addressed the global challenge of integrated station-tunnel construction. The rigid overhead catenary system ensures stable and reliable operation at the high speed of 160 km/h. The Fully Automatic Operation system represents the pinnacle of intelligence in urban rail transit, while the SuiTeng OS intelligent platform has fundamentally redefined the paradigm for metro operational management.

These technological innovations not only provide robust technical support for Line 18 itself but also lay a solid foundation for realizing the "One-Hour Commuting Circle" within the Guangdong-Hong Kong-Macao Greater Bay Area. Through the strategic implementation of "Four-Networks Integration" and cross-modal seamless transfers, Line 18 is actively reshaping the regional transportation landscape, catalyzing the GBA's transition from a collection of independently developing cities towards a functionally integrated, homogenized metropolitan region.

From the perspective of international vocational education, the technological advancements of Line 18 supply cutting-edge teaching content and practical training case studies. By systematically integrating Line 18's technical standards, operational procedures, and maintenance experience into vocational education and training systems, it becomes possible to cultivate technical skilled personnel who are equipped to meet the demands of an increasingly intelligent and high-speed rail transit industry. This process, in turn, facilitates the international export of Chinese rail transit technology, standards, and talent.

With the full completion and future extensions of Line 18, Guangzhou Metro is poised to continue its leadership role in driving technological innovation for China's urban rail transit sector. The line's achievements and the lessons learned from its implementation are set to contribute significantly more "Chinese Wisdom" and "Guangzhou Solutions" to the global development of urban rail transit systems.

Chapter 22: Guangzhou Metro Line 9: The World's First Metro Line Constructed on Karst Terrain

22.1 Project Overview and Geological Challenges

Guangzhou Metro Line 9 constitutes a vital component of Guangzhou's urban rail transit network. Traversing the Huadu and Baiyun Districts, the line stretches from Fei'eling in the west to Gaozeng Station on the northern extension of Metro Line 3 in the east. It serves key areas including the Huadu Automobile City, Guangzhou North Railway Station, Huadu District Government, and the Airport Business District. With a total length of 20.1 kilometers entirely constructed underground and comprising 11 stations, Line 9 commenced official operation on December 28, 2017. It holds dual significance as not only the first metro line within Huadu District but also a crucial transportation artery facilitating Guangzhou's strategic northern expansion and efficient development.

The most prominent characteristic of this line is its status as a world-first metro engineering project in karst geological conditions. Guangzhou Metro Line 9 is the first entirely underground line in China constructed within a water-rich, highly developed karst region. The geological conditions along the entire route are exceptionally complex, featuring adverse phenomena such as karst formations, fault zones, and water-rich sand layers. The karst development is particularly intense, with an average encounter rate for karst caves and soil cavities of 50%, reaching up to 70% at specific construction sites. During preliminary investigations in 2013, constructors drilled 3,353 boreholes, revealing a cavity encounter rate of 43.4%. This statistic signifies that nearly one out of every two boreholes intersected a karst cave or soil cavity. In areas of most intense karst development, such as Huadu Square and Guangzhou North Railway Station, the encounter rate reached 60%. Academician Qian Qihu of the Chinese Academy of Engineering described the line's geological situation with the phrase "exceptionally complex," noting, "The geology of Huadu District is abnormally complex. The entire 20-kilometer-long metro line is situated on karst terrain—this is a national first."

Table 22-1 Basic Information on Karst Geology of Guangzhou Metro Line 9

Item	Parameter / Data	Description / Elaboration
Total Line Length	20.1 km	Constructed entirely underground, navigating complex subsurface conditions throughout its entire route.
Number of Stations	11 Stations	Key stations include Fei'eling, Huadu Automobile City, and Guangzhou North Railway Station, connecting major urban hubs in Huadu and Baiyun Districts.
Karst Cavity Encounter Rate	Average: 50% Maximum: 70%	Based on 3,353 preliminary boreholes drilled along the alignment, an overall cavity (karst cave or soil cavity) encounter rate of 43.4% was recorded, indicating extreme geological complexity.
Maximum Cavity Height	27 meters	The largest single cavity encountered measured 27 meters in height, with another significant cavity reaching 19 meters, posing substantial engineering challenges for tunneling and ground stabilization.

Number of Underpassed Rivers	3 Rivers	The tunnel alignment crosses beneath three rivers: Tianma River, Tianmei River, and Qingshi River, requiring specialized water-proofing and ground settlement control measures.
Number of Underpassed High-Speed Railways	2 Crossings	The line passes twice beneath major national railway corridors: the Wuhan-Guangzhou High-Speed Railway and the Beijing-Guangzhou Railway, necessitating stringent deformation control to ensure the absolute safety of ongoing rail operations above.

The engineering challenges confronting Guangzhou Metro Line 9 extended far beyond its notorious karst geology. The alignment also necessitated underpassing multiple rivers, traversing beneath hundreds of residential and commercial buildings, and crossing critical infrastructure such as the Wuhan-Guangzhou High-Speed Railway and the Beijing-Guangzhou Railway. Progressing from west to east, the tunnel intersected six major fault zones—Xinghua Fault, Sanhua Fault, Tianmei Fault, Yayao Fault, Qingtan Fault, and Guangling Fault—each introducing potential risks of ground instability and water ingress. Furthermore, the route passed beneath three rivers—Tianma River, Tianmei River, and Qingshi River—and three expressways—Guangqing Expressway, Airport North Expressway, and Airport Expressway.

This convergence of highly adverse geological conditions and exceptionally sensitive surface structures rendered Line 9 the most challenging metro construction project in Guangzhou's history at its time. Owing to its unprecedented routing across such extensively developed karst terrain, the project was authoritatively recognized by experts as China's first metro line successfully constructed directly upon a soluble limestone (karst) rock foundation, setting a significant precedent for urban rail transit development under extreme geo-environmental constraints.

22.2 Key Technologies for Karst Investigation and Treatment

22.2.1 Refined Karst Investigation Technologies

Confronted with the exceptionally complex karst geological conditions, Guangzhou Metro Group adopted an ultra-high-density borehole investigation strategy. In 2010, Guangzhou Metro specifically convened a collaborative "expert diagnosis" of the geological risks for the Phase I project of Line 9, involving two Chinese Academy of Engineering academicians and over ten nationally renowned geological experts. Following the experts' "prescription," the constructors intensified and deepened the standard investigation protocol. Boreholes were drilled on an extremely dense grid: one hole every 3 meters around station areas, intensifying to every 2 meters in the most critical sections—a density more than ten times the standard requirement. Cumulatively, 3,353 boreholes were drilled along the entire line.

To accurately characterize the groundwater hydrology, Line 9 innovatively employed underground isotope testing technology. This technique analyzes variations in the composition of stable and radioactive isotopes within groundwater to trace the flow paths and velocities of karst groundwater, providing a scientific basis for predicting water inrush risks and formulating targeted karst treatment strategies. Concurrently, the construction team introduced Grey Fuzzy Theory for the first time in this context to analyze the developmental characteristics of karst caves, thereby establishing systematic principles and methodologies for tunnel karst treatment.

22.2.2 Karst Treatment and Reinforcement Technologies

For the identified subsurface cavities, the metro constructors implemented grouting methods to fill the voids. The total grout volume injected for karst cave and soil cavity treatment along the line reached approximately 430,000 cubic meters—sufficient to fill 171 standard Olympic-sized swimming pools. To address the diverse karst-related geotechnical challenges, the team developed several innovative technologies:

1. **Integrated Shield Tunneling Technology for Composite Strata under Water-Rich Karst Conditions:**Guangzhou Metro pioneered the establishment of principles and methods for treating karst areas and soil cavities. They developed a comprehensive suite of design and construction techniques for shield tunneling and

deep open-cut excavations under these complex geological constraints. This included creating emergency response technologies for integrated karst treatment in metro projects and innovating key methods for underground blasting to remove obstacles during shield tunneling, successfully resolving the technical challenges of construction in water-rich karst regions—a first-of-its-kind endeavor in China.

2. **Prefabricated Steel Sleeve Balanced Launching/Arrival Technology:**This technique replaced traditional end-head reinforcement methods. It involves assembling prefabricated steel sleeves and reaction frames directly against the end diaphragm wall on the station base slab. The sleeves are then filled with sand, slurry, or water and pressurized. This ensures the internal pressure within the sleeve reaches and maintains the design value. After breaching the diaphragm wall, the shield machine enters this sealed, pressurized environment within the sleeve. The stable pressure prevents a pressure differential from developing in the soil mass at the station head, thereby effectively preventing ground collapse or significant settlement.

3. **MJS (Metro Jet System) Omni-Directional High-Pressure Jet Grouting Method:** This technology was introduced for the first time in Guangdong Province specifically for horizontal ground reinforcement beneath existing railway lines, notably applied during the underpassing of the Wuhan-Guangzhou High-Speed Railway. The MJS method forms uniform reinforced columns through high-pressure jet grouting. Its unique feature is the integrated environmental monitoring and pressure control system, which effectively manages surface deformation, meeting the extremely stringent settlement control requirements essential for high-speed railway embankments.

Table 22-2: Key Technical Indicators for Karst Treatment in Guangzhou Metro Line 9

Treatment Technology	Application Scope	Technical Outcomes & Effectiveness
High-Pressure Grouting Filling	Treatment of karst caves and soil cavities along the entire line.	Cumulative grout volume reached430,000 cubic meters. Successfully reduced the cavity

		encounter rate and associated risks to within safe and manageable thresholds for construction, ensuring ground stability.
Steel Sleeve Reception Technology	Applied 8 times for shield machine launch/arrival at various stations along the line.	Effectively replaced traditional, more extensive end-head ground reinforcement. Resulted in reduced engineering costs and a smaller construction footprint/site enclosure area, minimizing surface disruption, especially in urban settings.
MJS (Metro Jet System) Method	Specifically deployed for the section underpassing the Wuhan-Guangzhou High-Speed Railway.	Utilizing 112 MJS piles, this method provided precise horizontal ground reinforcement. Achieved exceptional settlement control, limiting deformation to less than 5 mm, thereby ensuring the absolute safety and uninterrupted operation of the high-speed railway above.
Ground Freezing Method	Employed for the construction of cross-passages and specific challenging areas like Qingtang Station.	Enhanced construction efficiency and worker safety in water-bearing strata. Effectively resolved issues related to difficult ground conditions and limited site access ('land borrowing' difficulties) by creating a temporary, stable, water-tight frozen soil structure.
Trench Cutting Re-Mixing (TRM) /	Used for constructing Diaphragm Walls in sections encountering very hard rock	Overcame the significant technical challenge of efficiently and

Hydromill	formations.	accurately excavating trenches for diaphragm walls through extremely hard rock. Significantly improved construction efficiency and ensured the quality and integrity of the underground structural walls.

22.3 Breakthroughs in Shield Tunneling Engineering Technology

22.3.1 Innovations in Dual-Mode Shield Tunneling Technology

Guangzhou Metro Line 9 pioneered the development and application of a parallel-type dual-mode shield machine in China, effectively overcoming the challenge of traditional single-mode shields being ill-suited for the highly variable and complex geology of water-rich karst composite strata. This dual-mode shield machine achieved a significant advancement: the ability to switch between tunneling modes without requiring the assembly or disassembly of any components under specific conditions. This innovation endowed the shield machine with the high excavation efficiency characteristic of Earth Pressure Balance (EPB) shields, while simultaneously retaining the superior settlement control advantages of Slurry Shield (SS) machines.

During the construction of Line 9, a total of 16 shield machines were deployed, comprising 10 Slurry Shields, 4 Earth Pressure Balance Shields, and 2 Slurry/EPB Dual-Mode Shield machines. Practical construction experience revealed significant limitations for both single-mode types. While EPB shields demonstrated advantages in excavation efficiency and muck removal capability compared to SS machines, their settlement control was relatively weaker. Conversely, Slurry Shields offered significant advantages in precise ground settlement control but were less effective at handling large rock fragments, making them prone to slurry line blockages.

The dual-mode shield machine fully leveraged its strengths in strong adaptability, high risk resistance, and rapid excavation progress during construction. For instance, during the underpassing of the Tianma River—facing composite strata characterized by shallow overburden, fill containing boulders, developed karst caves, and soft upper layers overlying hard lower layers—the construction team successfully navigated the section by adopting a combined strategy: "advancing primarily in Slurry Mode, while switching to EPB Mode for boulder discharge." Utilizing methods like reverse circulation or standby pipe positive circulation combined with semi-reverse circulation, they effectively mitigated the risk of muck removal difficulties. This successful passage marked China's first shield tunneling project to underpass a river within a karst region.

22.3.2 Shield Control Technology under Complex Conditions

Addressing the unique "soft upper layer over hard lower layer" composite strata prevalent on Line 9, the construction team innovated shield axis control technologies for sharp curves within such geology. They also developed safe construction control protocols for shield tunneling at the interface between sandy soil layers and karst formations. A notable application was during the underpassing of the Guangqing Expressway section, which included entrance/exit ramps, the mainline, and toll plaza areas. Here, the tunnel cover above the mainline was only one tunnel diameter thick, primarily consisting of soft strata and localized soft-over-hard composites. The successful and safe underpassing was achieved through rigorous parameter control, comprehensive 3D management, thorough preparation for slurry mode operation, safe and rapid excavation utilizing EPB mode capabilities, and the implementation of detailed and feasible traffic diversion plans.

To tackle the challenges posed by hard rock and boulders to the shield machines, the metro constructors implemented a series of technical countermeasures. In the Xinnan–Fengxia shield section of Line 14's Knowledge City Line, the shield machine encountered an astonishing 33 uncharted boulders while grinding through a mere 500-meter tunnel stretch. The team conducted intensive supplementary surveys for boulders at a density of one borehole per meter and performed preemptive blasting on identified boulder clusters and bedrock protrusions. In another example, the Xiayuan–Nangang section of Line 13's first phase, spanning 4,936 meters, encountered localized rock with compressive strengths reaching 198 MPa—approximately 6 to 7 times harder than ordinary structural concrete. The constructors employed a "new tunnel boring energy-gathering smooth blasting technology utilizing hydraulic pressure" to effectively fracture the hard rock. Furthermore, based on site conditions, they utilized imported wear-resistant cutting tools and foam agents. These measures doubled the daily advance rate from less than 1.5 meters to approximately 3.0 meters, and significantly reduced cutterhead tool replacement frequency from once every 3 meters to once every 18 meters.

Table 22-2: Key Technical Indicators for Karst Treatment in Guangzhou Metro Line 9

Treatment Technology	**Application Scope**	**Technical Outcomes & Effectiveness**
High-Pressure Grouting Filling	Treatment of karst caves and soil cavities along the entire alignment	Cumulative grout volume reached 430,000 m³, successfully reducing cavity encounter rates to safe construction thresholds
Steel Sleeve Reception Technology	Applied at 8 shield launching/arrival locations along the line	Effectively replaced conventional end reinforcement, reducing construction costs by 15% and minimizing construction footprint area by 30%
MJS (Metro Jet System) Method	Section undercrossing Wuhan-Guangzhou High-Speed Railway	Installation of 112 MJS piles achieved exceptional settlement control with maximum deformation limited to $\leq$5 mm
Ground Freezing Method	Cross passages and Qingtang Station construction	Enhanced construction efficiency by 40% and resolved site access limitations in constrained urban environments
Trench Cutting Re-Mixing (TRM) / Hydromill	Diaphragm wall construction in hard rock formations	Successfully overcame trenching challenges in rock strengths up to 80 MPa, improving construction

efficiency by 60%

22.3 Technological Breakthroughs in Shield Tunneling Engineering

22.3.1 Innovation in Dual-Mode Shield Tunneling Technology

Guangzhou Metro Line 9 pioneered the development and application of a parallel dual-mode shield system in China, effectively overcoming the limitations of conventional single-mode shields in adapting to the complex and variable geological conditions of water-rich karst composite strata. The dual-mode shield machine achieved a significant technological advancement: the ability to switch between tunneling modes without requiring component assembly or disassembly under specific conditions. This innovation combines the high excavation efficiency characteristic of Earth Pressure Balance (EPB) shields with the superior settlement control advantages of Slurry Shield systems.

During the construction of Line 9, a total of 16 shield machines were deployed, comprising 10 Slurry Shields, 4 Earth Pressure Balance Shields, and 2 Slurry/EPB Dual-Mode Shield machines. Field experience demonstrated significant limitations of single-mode systems. While EPB shields showed advantages in excavation efficiency and muck removal capability compared to Slurry Shields, they exhibited relatively weaker settlement control. Conversely, Slurry Shields offered superior precision in ground settlement control but demonstrated insufficient capacity for handling large rock fragments, frequently causing slurry discharge pipe blockages.

The dual-mode shield machine fully leveraged its advantages of strong adaptability, high risk resistance, and rapid excavation progress during construction. A notable application was the undercrossing of Tianma River, where the tunnel encountered complex strata conditions including shallow overburden, fill material with boulders, developed karst caves, and soft-over-hard geological interfaces. The construction team successfully implemented a combined strategy of "primary advancement in Slurry Mode with periodic switching to EPB Mode for boulder discharge." By employing reverse circulation or standby pipe positive circulation combined with semi-reverse circulation methods, they effectively mitigated muck removal difficulties and safely completed the crossing, establishing China's first successful case of shield tunneling beneath a river in a karst region.

22.3.2 Shield Control Technology Under Complex Conditions

To address the unique "soft-over-hard" composite strata characteristic of Line 9, the construction team developed innovative shield alignment control technologies for sharp curves in such conditions and established safe construction protocols for tunneling at interfaces between sandy soil layers and karst formations. During the undercrossing of the Guangqing Expressway section—including toll plaza entrance/exit ramps, mainline sections, and the toll station area—where the tunnel cover above the mainline measured only one tunnel diameter, primarily consisting of soft strata with localized soft-over-hard composites, the team successfully completed the undercrossing through rigorous parameter control, comprehensive 3D management, thorough preparation for slurry mode operation, safe rapid excavation using EPB mode capabilities, and implementation of detailed traffic diversion plans.

To overcome challenges posed by hard rock and boulders to shield operations, metro constructors implemented multiple technical countermeasures. In the Xinnan-Fengxia shield section of Line 14's Knowledge City Line, the shield machine encountered 33 uncharted boulders while advancing through a mere 500-meter tunnel stretch. The team conducted intensive supplementary surveys at one-borehole-per-meter density and performed preemptive blasting of identified boulder clusters and bedrock protrusions. In the Xiayuan-Nangang section of Line 13's initial phase, spanning 4,936 meters with localized rock strength reaching 198 MPa—approximately 6-7 times harder than conventional structural concrete—constructors implemented "new energy-concentrated hydraulic smooth blasting technology for tunnel excavation" to effectively fracture hard rock. Combined with imported wear-resistant cutters and foam agents tailored to site conditions, these measures doubled the daily advance rate from less than 1.5 meters to approximately 3.0 meters, while extending cutter replacement intervals from every 3 meters to every 18 meters.

Table 22-3 Performance Comparison: Dual-Mode Shield vs. Conventional Shields

Performance Indicator	Earth Pressure Balance (EPB) Shield	Slurry Pressure Balance (SPB) Shield	Dual-Mode Shield (EPB/SPB)
Advance Rate	High	Moderate	High (adjustable based on ground conditions)
Settlement Control	Relatively Poor	Excellent	Excellent
Adaptability to Karst Strata	Low	Moderate	High
Initial Equipment Investment	Low	Moderate	Relatively High
Operational Cost	Low	Relatively High	Moderate
Mode Switching Time	-		15-32 mode switches performed without component assembly/disassembly

22.3.3 Special Construction Methods and Material Technologies

Line 9 implemented ground freezing technology at multiple critical nodes. Due to constraints such as limited site availability and existing underground utilities, the project innovatively adopted this method at the No. 2 cross passage between Huadu Automobile City Station and Guangzhou North Railway Station, the cross passage in the Lianqing section, and at Qingtang Station, significantly enhancing construction efficiency. The ground freezing method creates a temporary frozen curtain underground, effectively sealing off groundwater and providing a dry, stable working environment for tunnel construction.

In terms of support technology, Line 9 employed comprehensive construction techniques for diaphragm walls in hard rock. This included utilizing trench cutter units, imported rotary drilling rigs combined with impact hammers, and surface drilling with pre-splitting blasting to overcome challenges associated with trench excavation in hard rock. To address varying ground conditions, the team selected different types of shield machines, including 10 slurry shield machines, 4 dual-screw earth pressure balance (EPB) shield machines, and 2 slurry/earth pressure dual-mode shield machines. The dual-screw EPB shield machines were specifically designed for hard rock and soft-over-hard composite strata, utilizing dual screw conveyors to achieve efficient muck discharge and precise pressure control.

22.4 Key Technologies for Underpassing High-Speed Railways

22.4.1 Technical Challenges and Control Standards

The tunnel section between Guangzhou North Station and Huacheng Road Station on Guangzhou Metro Line 9 successfully underpasses both the Wuhan-Guangzhou High-Speed Railway and the Beijing-Guangzhou Railway, making it the first metro shield tunnel in China to pass beneath a high-speed railway embankment. This project presented immense challenges: the shield tunnel section needed to underpass four tracks and the platform canopy of the Wuhan-Guangzhou High-Speed Railway (designed for 350 km/h), as well as six tracks of the Beijing-Guangzhou Railway (160 km/h). The closest distance between the tunnel's external profile and the pile foundation of the high-speed railway station's platform canopy was a mere 1.2 meters. With the metro shield tunnel situated "dangerously close" to the high-speed railway embankment, any slight negligence during construction could have adversely affected these two critical national railway arteries.

Particularly daunting was the fact that there were no domestic precedents for metro tunnels underpassing high-speed railway embankments and ballastless tracks, leaving no established experience to draw upon. Construction in the karst area of this section, characterized by shallow overburden with sand layers overlying limestone, was especially prone to settlement incidents, potentially failing to meet the stringent settlement control requirements of the high-speed railway. Starting in 2010, the design scheme for underpassing the Wuhan-Guangzhou High-Speed Railway underwent six years of research and six rounds of optimization and adjustment. Subjected to rigorous review by three Chinese Academy of Engineering academicians (Qian Qihu, Lu Yaoru, Wang Mengshu) and design master Shi Yuxin, among other experts, the final settlement control requirement for the metro construction's impact on the high-speed railway was set at a maximum of 5 millimeters. This threshold was far stricter than the 30-mm ground settlement control value specified for shield tunnels in relevant national standards.

22.4.2 Technological Breakthroughs and Implementation Outcomes

To meet these extremely stringent settlement requirements, Guangzhou Metro and the participating parties developed a methodology involving "first reinforcing the railway embankment, then having the shield tunnel through the reinforced soil mass beneath the railway." They innovatively adopted the "Metro Jet System (MJS)" method, completing a total of 112 MJS piles within the approximately 100-meter-long section of strata beneath the Wuhan-Guangzhou High-Speed Railway and Beijing-Guangzhou Railway where the shield tunnel passed. The MJS method, through its unique environmental sensing and pressure control technology, effectively managed surface deformation, satisfying the demanding settlement control criteria for the high-speed railway embankment.

Concurrently, the construction team established an automated monitoring software platform for the first time, utilizing information technology to address construction safety concerns. This platform provided real-time monitoring of shield tunneling parameters and railway track deformation, allowing for immediate adjustment of tunneling parameters upon detecting any anomalies, thereby ensuring settlements remained consistently within the permissible limits. Ultimately, the tunnel section between Guangzhou North Station and Huacheng Road Station on Line 9 successfully underpassed both the Wuhan-Guangzhou and Beijing-Guangzhou railways, achieving the demanding targets of "settlement not exceeding 5 mm and absolutely no heave" to a high standard.

This technical achievement was summarized as the "Key Technology for Underpassing Ballastless Track Embankments of 350 km/h High-Speed Railways." It marked the first successful instance in China where shield tunneling beneath a high-speed railway embankment achieved strictly controlled deformation without requiring speed restrictions on the railway, filling a significant gap in the domestic industry. This technology provides crucial reference for subsequent similar projects, offering essential technical support for coordinating underground space development with the protection of railway facilities, particularly in increasingly dense high-speed railway networks within urbanizing areas.

22.5 Intelligent Operation and Maintenance and Electromechanical Innovation Technologies

22.5.1 Intelligent Inspection and Operation & Maintenance System

During the operation and maintenance phase, Guangzhou Metro Line 9 has adopted several intelligent and innovative technologies, the most representative being the industry's first intelligent inspection robot for urban rail transit vehicles. This robot commenced trials in April 2020 at the Qishan Depot of Line 9 and passed the review by the Technical Equipment Committee of the China Association of Metros in January 2021. Review experts unanimously recognized that it possesses fully independent intellectual property rights, fills a gap in intelligent inspection for urban rail vehicles both domestically and internationally, and achieves an internationally leading level within the urban rail transit industry.

The intelligent inspection robot utilizes advanced technologies including SLAM autonomous navigation, multi-degree-of-freedom flexible collaborative robotic arm control, high-definition optical imaging, image pattern recognition, and AI deep learning algorithms. It is capable of performing 2D/3D high-definition imaging of key inspection points beneath the train, intelligently judging abnormal states of key undercarriage components, achieving automatic transfer across multiple tracks, and performing automated detection, intelligent analysis, and fault diagnosis for vehicle inspection tasks. As of now, this intelligent inspection robot has cumulatively inspected 2,241 trains, achieving a fault recognition rate of 98.1%.

Complementing this is the 360° Train Image Online Detection System. This system employs high-definition line-scan imaging with near-infrared laser fill-light technology, image recognition technology, and deep learning technology. It covers the automated detection and identification of visually inspectable key components across the entire 360° range of the train, including the roof, sides, and underbody. Fixed-installed alongside the tracks, the 360° system can scan and inspect components across the entire "body" of a moving train, quickly and efficiently, covering over 9,000 inspection items. Meanwhile, the flexible intelligent inspection robot is responsible for detailed scanning inspections of hard-to-reach areas and "nooks and crannies" of the train, covering 2,478 inspection items.

22.5.2 Intelligent Electromechanical and Energy-Saving Technologies

Guangzhou Metro Line 9 pioneered an intelligent system for urban rail transit electromechanical systems. It innovatively adopted evaporative cooling direct expansion technology, incorporated regenerative braking energy inverter feedback devices in the power supply system, implemented a centralized UPS power supply scheme, and deployed an intelligent lighting control system. By utilizing energy-efficient and environmentally friendly high-performance dynamic compaction equipment and applying key foundation treatment technologies, it broke through technical bottlenecks related to construction period, investment, energy saving, and environmental protection encountered when dealing with complex foundations. This addressed the technical challenges of "Four Savings and One Environmental Protection" in engineering construction, resulting in annual electricity savings of 1.46 million kWh and cost savings of 1.168 million RMB.

Regarding the power supply system, Line 9 is equipped with regenerative braking energy inverter feedback devices. These devices recover the energy generated during train braking and feed it back into the power grid for use by other trains or equipment, significantly reducing the line's operational energy consumption. The centralized UPS power supply scheme provides unified uninterrupted power supply for critical equipment, enhancing system reliability while reducing the space occupation and maintenance costs associated with decentralized arrangements. The intelligent lighting control system automatically adjusts station lighting intensity based on passenger flow and natural light conditions, achieving an optimal balance between energy savings and passenger comfort.

22.5.3 BIM and Information Management Technology

The Line 9 project pioneered key technologies for information-based control in urban rail transit engineering construction. It marked the first establishment of key technologies for safety risk prevention and control in urban rail transit engineering, creating an intelligent safety risk prevention and control information platform. It also developed a refined management system for civil engineering, achieving information-based services across all levels. The project developed an integrated metro project management platform, fully implemented the "Four Modernizations" management principle, established a BIM information technology management system, and formed a cross-enterprise, refined engineering project management system. This enables visual, traceable, and accountable process management for construction operation procedures across different disciplines.

BIM technology played a vital role in the construction of Line 9, especially in complex project segments like karst treatment and underpassing high-speed railways. Through pre-construction simulation and scheme optimization using BIM models, numerous potential problems were identified and resolved in advance, reducing construction changes and rework. Simultaneously, integrating BIM technology with the karst geological database and real-time monitoring systems formed a digital twin model, providing accurate data support for operation and maintenance.

Table 22-4 Application Effectiveness of the Intelligent Inspection System on Guangzhou Metro Line 9

Metric	**Traditional Manual Inspection**	**Intelligent Inspection System**	**Improvement Effect**
Inspection Efficiency	2 hours/train	0.5 hours/train	75% Increase
Number of Inspection Items	~5,000 items	>11,000 items	120% Increase
Fault Identification Accuracy	~5,000 items	98.1%	+8.1 Percentage Points
Night Shift Personnel Count	8 persons/shift	5 persons/shift	37.5% Reduction
Data Traceability	Paper-based recording	Digital recording &	Significantly Enhanced

		storage	
Mean Time to Fault Detection	During next scheduled maintenance	Real-time detection	Substantially Advanced

22.6 Significance for International Vocational Education

22.6.1 Technological Innovation and Knowledge System

The construction achievements of Guangzhou Metro Line 9 are manifested not only in the physical project but also in providing a complete technical solution and knowledge system for urban rail transit construction in international karst regions. The "Overall Requirements for Karst Treatment in the Guangzhou Metro Line 9 Project," compiled based on the experience from Line 9, has formed an industry technical standard, providing a technical basis for the design and construction of rail transit in limestone areas, such as the northern extension of Guangzhou Metro Line 8.

The six "domestically pioneering" technological achievements developed during the construction of Line 9, covering areas like karst geological treatment, shield tunneling engineering, underpassing high-speed railways, and electromechanical systems, constitute a comprehensive technological system for metro construction in karst regions. These achievements, systematically compiled, have been published in the monograph "Key Technologies and Practices in Civil Engineering for Guangzhou Metro Line 9," serving as an authoritative textbook for international vocational education.

In terms of vocational education, the successful experience of Line 9 demonstrates that metro construction in karst areas must emphasize three core principles: prioritizing geological investigation, ensuring precise technical solutions, and implementing strict process control. Through detailed investigation and scientific analysis, the patterns of karst development can be accurately understood; targeted technical solutions address various complex geological challenges; and strict process control with real-time monitoring ensures project quality and safety. This philosophy is worthy of promotion in the field of international vocational education.

22.6.2 Implications for International Urban Rail Transit Construction

As the world's first metro line constructed under geological conditions of shallow overburden and intensely developed karst, the successful practice of Guangzhou Metro Line 9 offers significant insights for similar urban rail transit projects internationally:

Firstly, Line 9 proves that constructing high-quality metro projects in extremely complex karst geological conditions is feasible. Through technological innovation and refined management, adverse conditions like karst, fault zones, and water-rich sand layers can be overcome, enabling safe and efficient metro construction. This provides confidence and a technical pathway for international cities with similar geological challenges.

Secondly, the dual-mode shield tunneling technology developed for Line 9 offers a new equipment solution for metro construction in composite strata. Especially in areas with frequent geological variations, the flexibility and adaptability of the dual-mode shield provide distinct advantages. This technology has already generated widespread influence in international tunneling engineering.

Thirdly, the technology and control standards established for underpassing high-speed railways on Line 9 set a benchmark for engineering where urban rail transit crosses high-speed railway lines. As global high-speed rail networks and urban rail transit systems become increasingly dense, such crossings will become more common, making the experience of Line 9 highly valuable for reference.

Finally, the "human-machine combination" vehicle maintenance model created for Line 9 offers a new approach for urban rail transit operation and maintenance. By integrating the intelligent inspection robot with the 360° image detection system, inspection efficiency and quality are improved while human resource allocation is optimized. This holds particular reference value for developed countries with high labor costs.

The construction experience of Guangzhou Metro Line 9 has been disseminated globally through international academic conferences, professional journal papers, and technical exchanges. Several of its technological achievements have already been adopted by urban rail transit projects in karst-developed areas such as Bangkok and

Kuala Lumpur. In the future, with the deepening of infrastructure cooperation under the Belt and Road Initiative, the value of the technological innovations and vocational education aspects of Guangzhou Metro Line 9 will be further demonstrated on the international stage.

Chapter 23: The Nation's First Three-Story Underground Prefabricated Station

23.1 Project Overview and Technological Innovation Value

Guangzhou Metro Line 11, as Guangzhou's first major metro loop line, spans 44.2 kilometers, traversing five central districts: Haizhu, Tianhe, Baiyun, Yuexiu, and Liwan. It interconnects major transportation hubs including Guangzhou Railway Station, Guangzhou East Railway Station, and the Pearl River Delta Intercity rail network, rightfully earning its title as the "King of Interchanges." On this technology-intensive line, Shangyong Park Station stands out as a milestone project in China's urban rail transit construction history, being the nation's first three-story underground prefabricated station and the first prefabricated station in South China.

Located within Shangyong Park, northwest of the intersection of Guangzhou Avenue and Xinjiao Road in Haizhu District, Shangyong Park Station generally follows an east-west orientation. The station has a total length of 221.7 meters and a standard width of 22.3 meters. Its construction broke through the technical bottlenecks of traditional metro station building, pioneering the application of prefabricated technology in deep underground spaces.

Situated in a bustling area of Haizhu District and located within a rock stratum fracture and fragmentation zone, the station posed extremely high demands for construction organization, safety, and quality control. To address these challenges, the project team empowered the construction process with smart digital technologies. They comprehensively adopted an integrated information management platform, a hidden danger investigation system, video surveillance, and other multi-functional digital-intelligent equipment. This was complemented by the implementation of advanced production facilities such as an intelligent steel reinforcement processing plant and intelligent gantry cranes, comprehensively enhancing the level of safe, civilized, and standardized construction.

Table 23-1 Basic Engineering Parameters of Shangyong Park Station

Item	**Parameter / Specification**	**Remarks / Elaboration**
Station Type	Three-Story Underground Prefabricated Station	Represents China's first application of this structural typology in deep underground metro station construction, demonstrating significant innovation in construction methodology.
Station Length	221.7 meters	The total station length, with a 105-meter section constructed using prefabricated elements, establishing the core prefabricated zone.
Standard Section Width	22.3 meters	Refers to the typical cross-sectional width of the station structure in its standard configuration.

Excavation Depth	Approx. 30 meters	The depth of the main excavation, with localized areas potentially requiring deeper excavation depending on sub-surface conditions and structural requirements.
Prefabricated Component Types	8 Categories, 18 Precast Segments	Includes major structural elements such as side walls and center columns, showcasing the diversity and complexity of the prefabricated system deployed.
Overall Prefabrication Rate	63.6%	Indicates that a substantial majority of the station's structural components were manufactured off-site and assembled on-site, highlighting a high degree of industrialized construction.
Construction Method	Open-Cut Top-Down Method + Prefabricated Assembly Method	Combines traditional cut-and-cover excavation sequenced from the top down with the innovative assembly of prefabricated structural components, optimizing construction workflow and timeline.

23.2 Design Philosophy and Technical Characteristics of the Prefabricated Station

23.2.1 Design Philosophy and Principles

The design philosophy for the prefabricated metro station originates from an industrialized construction ideology, aiming to transform traditional cast-in-situ wet works into a dry-assembly model characterized by factory production and on-site assembly. This transition not only enhances construction efficiency but also significantly improves project quality and environmental performance. The design of Shangyong Park Station on Guangzhou Metro Line 11 fully considered factors such as structural mechanics, station service functionality, and project cost, adopting a novel prefabricated metro station structure based on dry connections.

Regarding the cross-sectional design, the station's internal clear width, arch rise radius of the roof slab, and structural clear height of the concourse level are 18.3 meters, 22.0 meters, and 6.6 meters, respectively. This optimized cross-sectional configuration carefully balances passenger flow space, equipment layout requirements, and structural mechanical performance. Based on the segmentation principles for prefabricated stations, the design team finalized a segmentation scheme suitable for the internal bracing system. By optimizing the configuration of the prefabricated formwork to increase its versatility, they enhanced the adaptability of the design solution.

23.2.2 Structural System and Segmentation Technology

The prefabricated structural system of Shangyong Park Station employs multiple components combined ring-wise, namely the base slab component, bottom corner component, sidewall component, middle slab component, top corner component, and roof slab component. The edges of each component are equipped with mutually matching interlocking male-female structures, and the interfaces between adjacent components are filled with a sealing system. Furthermore, the edges of the components feature longitudinally extending dovetail grooves; the grooves of adjacent components align opposite each other, housing connecting elements installed within them to generate mutually opposing clamping forces on the adjacent components.

This structural design ensures that the connecting elements apply opposing clamping forces to adjacent components, guaranteeing that the sealing system at the interface provides superior sealing performance. Utilizing a unified connecting element instead of multiple bolt structures offers greater moment resistance and more uniform force transfer, resulting in a more robust clamping force between adjacent components. This satisfies high-performance waterproofing requirements, simplifies on-site operations, and increases construction efficiency.

Table 23-1 Types and Functions of Prefabricated Components in the Prefabricated Station

Component Type	Quantity	Primary Function	Installation Characteristics
Precast Side Wall	Multiple Units	Resists lateral earth and water	Requires precise micro-adjustment

		pressures, forms the primary vertical enclosure of the station.	during placement and temporary bracing to prevent sway or displacement until fully connected and backfilled.
Center Column	Multiple Units	Provides vertical support for the middle slab and roof slab, transferring loads to the foundation.	Demands extremely high verticality control, with alignment tolerances typically within 0.2%, often ensured by laser guidance systems.
Middle Slab	Multiple Units	Structurally separates the station concourse level from the platform level, providing intermediate support and diaphragm action.	Involves lifting and positioning large-area panels; requires precise alignment with supporting columns and walls for proper load transfer.
Roof Slab	Multiple Units	Forms the top structural enclosure of the station, supporting overburden loads and any surface traffic or structures.	High-precision installation is critical due to significant structural loads and stringent, long-term waterproofing requirements at joints.
Longitudinal Beam	Multiple Units	Enhances the overall transverse stiffness and integrity of the station structure, connecting and stabilizing adjacent columns.	Requires precise interface connection with column heads or corbels to ensure effective moment transfer and structural continuity.
Bottom Corner Component	Multiple Units	Connects the base slab to the sidewalls, forming a monolithic corner that effectively transfers shear forces.	Critical for ensuring the strength and stability of the wall-to-base connection node, often involving complex reinforcement detailing.

23.3 Key Construction Technologies and Process Innovations

23.3.1 High-Precision Diaphragm Wall Construction Technology

As a three-story underground station, Shangyong Park Station required a deep excavation pit, placing extremely high demands on the quality of the retaining structure. The project team employed the trench cutter (hydromill) technique for panel excavation. This was combined with the machine's own deviation correction system and ultrasonic testing methods to control the verticality and surface regularity of the excavated panels. To address challenges posed by the mudstone strata encountered in the ground conditions, the team developed a combined tool approach utilizing bits and disc cutters, which successfully resolved the issue of cutterhead clogging ("boulder-ing") in the mudstone.

More innovatively, the team developed a construction technology where the station's diaphragm wall also functioned as the permanent structural sidewall. This effectively improved the construction quality for this integrated "diaphragm wall as permanent structure" system. This technique broke from the conventional practice of separating the temporary retaining structure from the permanent main structure in metro station construction. By achieving this "two-walls-in-one" integration, it saved materials, enhanced construction efficiency, and shortened the project schedule.

23.3.2 Precision Installation Technology for Steel Tubular Columns

In the three-story underground station structure, the vertical alignment accuracy of the steel tubular columns directly impacts the structural load-bearing performance and safety. For the Shangyong Park Station project, a "Full-Rotation Drilling Rig + Flange Plate Positioning" method was adopted to control the verticality of the steel columns. The innovatively developed construction technology for steel tubular columns in prefabricated metro stations enabled precise control of the column verticality to an exceptionally strict standard of 0.2%, achieving an industry-leading level.

This technology utilizes full-rotation drilling rigs for precise boring, followed by positioning and fixing the steel columns using specially designed flange plates, ensuring extremely high vertical alignment is maintained during installation. As the key vertical load-bearing components of the station structure, the precise installation of these steel tubular columns laid a solid foundation for the subsequent hoisting and assembly of the prefabricated components.

23.3.3 Intelligent Hoisting Technology for Prefabricated Components

Facing the challenges of difficult hoisting for large prefabricated components and high requirements for positioning accuracy, the project team developed an intelligent hoisting device specifically for prefabricated metro station components. This device features precise micro-adjustment, electrical anti-sway, and fixed-point operation functions, enabling the efficient and precise installation of large-area prefabricated slabs in the assembly of the metro station.

The intelligent hoisting system employs a combination of real-time measurement and dynamic adjustment technologies to accurately control the placement and fixing of prefabricated components, ensuring accurate structural positioning. To guarantee the efficient progress of the prefabricated component assembly work, the project department implemented "whole-chain" refined management, strictly enforced a "three-inspection system" and a dual acceptance procedure (pre-dispatch and post-delivery), and utilized a mobile video surveillance system to track prefabricated components throughout the process, ensuring dimensional accuracy for every piece.

23.4 Intelligent Construction and Quality Control

23.4.1 Smart Site Management System

The Shangyong Park Station project comprehensively adopted an integrated information management platform, a hidden danger investigation system, video surveillance, and other multi-functional digital-intelligent devices, establishing a comprehensive intelligent construction management system. This system integrates multiple modules including schedule management, quality management, safety management, and material management, achieving visual, digital, and intelligent management of the entire construction process.

Particularly in the management of prefabricated components, the project team deeply integrated BIM technology with construction management. Using 3D modeling to simulate the construction sequence of the support system, they optimized the design of the joints between prefabricated components and the main structure, reducing interference between different work sequences. Simultaneously, relying on the smart site platform, the project team performed real-time measurement and verification of prefabricated component positions, providing a basis for scientific decision-making and helping to increase construction efficiency by over 20%.

23.4.2 Precise Control via the Axial Force Servo System

Addressing the challenge of deformation control in deep excavation construction, the project introduced an Axial Force Servo System, equipping the metro station's steel struts with a "most powerful brain." This system integrates numerical control hydraulic technology, automated monitoring technology, and IoT technology. It enables real-time monitoring of strut axial force and excavation deformation while simultaneously allowing for timely adjustment of the strut axial force based on minute deformation data, providing round-the-clock safety protection for the excavation's retaining structure.

Compared to conventional struts, which react passively after diaphragm wall deformation occurs, the axial force servo struts actively apply pressure to the diaphragm wall. This effectively ensures safety during station excavation. The system's detection accuracy reaches the millimeter level, allowing the steel struts to extend and retract freely, akin to a human arm, thereby making the deformation and stress state of the excavation structure controllable. This technology is particularly crucial for construction sites in urban centers adjacent to existing buildings, as it effectively controls excavation deformation and ensures the safety of the surrounding environment.

23.4.3 Whole-Chain Refined Quality Management

The Shangyong Park Station project implemented a strict "Whole-Chain" refined quality management approach. A comprehensive quality control system was established for every stage, from the production and transportation of prefabricated components to their on-site hoisting and assembly. During the production phase, strict control was maintained over the quality of raw materials and embedded parts for the prefabricated components, with random sampling and testing of relevant indicators conducted for each batch to achieve full traceability.

On the construction site, the team strictly implemented the "three-inspection system" and the dual acceptance procedure (pre-dispatch and post-delivery), ensuring every single prefabricated component conformed to the design requirements. Concurrently, safety precautions for lifting operations were continuously reinforced by establishing lifting operation warning zones, conducting pre-shift safety briefings, and organizing emergency drills, thereby building multiple layers of safety defense. This comprehensive quality management system ensured the construction quality and safety of the prefabricated station.

Table 23-2 Construction Indicators: Prefabricated Station vs. Traditional Cast-in-Situ Station

Metric	**Traditional Cast-in-Situ Station**	**Prefabricated Station**	**Improvement Effect**
Construction Period	Long Duration	Reduced by over 30%	Significantly Shortened, primarily due to parallel off-site manufacturing and streamlined on-site assembly processes.
On-site Workforce	Large Crew Size	Reduced by over 40%	Substantial Labor Savings, shifting labor demands from on-site to controlled factory environments.
Construction Noise & Dust	High Levels	Significantly Lowered	Enhanced Environmental Performance, minimizing disruption to adjacent communities and improving working conditions.
Quality Control	Highly Dependent on Variable Site Conditions	Factory-based Production	Superior and Consistent Quality, achieved through standardized manufacturing processes under controlled conditions.
Material Waste	Relatively High	Markedly Reduced	Improved Sustainability & Resource Efficiency, aligning with green building principles and reducing landfill burden.
Impact on Surrounding	Substantial Disruption	Noticeably Diminished	Pronounced Social Benefits, including less traffic interference, noise pollution, and overall

Environment	neighborhood disturbance.

23.5 Technological Innovation and Achievement Value

23.5.1 Technical Patents and Standard Development

Relying on the research, development, and application of prefabricated technology at Shangyong Park Station, the project has secured a total of 8 invention patents and 10 utility model patents. It has also been awarded numerous prestigious honors, including the Science and Technology Progress Award from the China Construction Industry Association, the Science and Technology Award from the Guangdong Provincial Railway Society, and the Science and Technology Award from China Railway Group Limited. These technological innovations cover multiple stages of prefabricated station development, including design, manufacturing, and construction, forming a complete independent intellectual property system.

Based on the practical experience gained from Shangyong Park Station, the project team actively participated in the formulation of 4 relevant industry specifications and acceptance standards. This process transformed the innovative achievements and successful experiences from the project into shareable technical specifications for the industry, effectively filling several technical gaps in the field of prefabricated metro station construction within China and laying a solid foundation for the further promotion and application of this technology. These standards provide a technical basis for subsequent prefabricated metro station construction, driving technological progress within the industry.

23.5.2 Comprehensive Benefit Analysis

As China's first three-story underground prefabricated station, the construction and operation of Shangyong Park Station have yielded significant comprehensive benefits. Economically, prefabricated construction reduced on-site wet works, improved construction efficiency, shortened the project schedule, and lowered the overall cost. Although the initial investment might be slightly higher than traditional construction methods, the whole-life-cycle cost demonstrates a clear advantage.

Regarding social benefits, prefabricated construction significantly reduced the disruption to the surrounding environment, decreased emissions of pollutants like noise and dust, and enhanced the green and environmental standards of the construction industry. Simultaneously, as a demonstration project, Shangyong Park Station has trained a large number of technical professionals in prefabricated building technology for the industry, promoting industrial upgrading and talent development.

In terms of environmental benefits, prefabricated construction reduced the generation of on-site construction waste, minimized material wastage, and aligned with the principles of green construction and sustainable development. Through factory production, the volume of on-site work was substantially reduced, effectively controlling the impact on surrounding traffic and the environment.

23.6 Significance for International Vocational Education

23.6.1 Technological Innovation and Knowledge System Enhancement

As China's first three-story underground prefabricated station, Guangzhou Metro Line 11's Shangyong Park Station provides significant technical reference and engineering demonstration value for international urban rail transit construction. Its successful practice demonstrates that employing prefabricated technology for constructing deep underground spaces is feasible and efficient under complex urban environmental conditions, offering new approaches for rail transit construction in countries worldwide facing similar geological and urban challenges.

The six "domestically pioneering" technological achievements developed through Shangyong Park Station, covering areas such as prefabricated station design, precast component manufacturing, precision construction installation, and intelligent management, constitute a complete technological system for prefabricated metro station construction. These achievements, systematically consolidated, have formed a set of promotable and replicable technical standards and management models, serving as authoritative teaching material for international vocational education.

In vocational education, the successful experience of Shangyong Park Station indicates that modern urban rail transit construction needs to emphasize four main principles: design standardization, production industrialization, construction assembly, and management informatization. Standardized design improves component universality, industrialized production ensures component quality, assembled construction enhances project efficiency, and information management enables whole-process control. This philosophy is worthy of promotion within the field of international vocational education.

23.6.2 Implications for International Urban Rail Transit Construction

The successful of Guangzhou Metro Line 11's Shangyong Park Station, as China's first three-story underground prefabricated station, offers important 启示 for international urban rail transit construction:

Firstly, Shangyong Park Station proves that constructing high-quality prefabricated metro stations is feasible in bustling urban areas and complex geological conditions. Through technological innovation and refined management, adverse geological conditions like rock fracture zones can be overcome, enabling safe and efficient metro construction. This provides confidence and a technical pathway for international cities with similar geological challenges.

Secondly, the trench cutter (hydromill) technology, precision installation technology for steel tubular columns, and intelligent hoisting technology for prefabricated components developed at Shangyong Park Station provide a complete set of technical solutions for prefabricated metro station construction. Particularly in areas with frequently varying ground conditions, the flexibility and adaptability of these technologies offer distinct advantages and have already generated widespread influence in international tunneling engineering.

Thirdly, the "factory prefabrication + on-site assembly" construction model established at Shangyong Park Station offers a new approach for urban rail transit construction. Through standardized design and industrialized production, it enhances project quality while shortening the on-site construction period, holding particular reference value for rail transit construction in urban city centers.

Finally, the Axial Force Servo System and BIM information technology applied at Shangyong Park Station provide a paradigm for the intelligent construction of urban rail transit. As global urbanization accelerates and the challenges facing urban rail transit construction become increasingly complex, the smart construction experience of Shangyong Park Station holds significant reference value.

The construction experience of Guangzhou Metro Line 11's Shangyong Park Station has been disseminated globally through international academic conferences, professional journal papers, and technical exchanges. Several of its technological achievements have already been adopted by rail transit projects in developed cities like Singapore and Hong Kong. In the future, with the deepening of infrastructure cooperation under the Belt and Road Initiative, the value of the technological innovations and vocational education aspects of Guangzhou Metro's prefabricated stations will be further demonstrated on the international stage.

Part VI: Vocational Education Knowledge Sharing

Chapter 24: International Vocational Education Cooperation Models and Practical Cases

24.1 Strategic Role of Guangzhou Metro's Vocational Education Export

As a leading urban rail transit enterprise in China, Guangzhou Metro Group has actively implemented its "Going Global" strategy in recent years, participating in the global rail transit market through diversified models. As of 2025, Guangzhou Metro Group's technology and knowledge transfer services have extended to over 100 cities worldwide, forming comprehensive international service capabilities across the entire industry chain, encompassing planning and design, engineering construction, operation management, and talent cultivation. This outstanding achievement signifies Guangzhou Metro's successful transformation from a single local operator into a comprehensive rail transit service provider with global influence.

The implementation of Guangzhou Metro's internationalization strategy is characterized by a systematic, multi-dimensional, and comprehensive approach. In terms of strategic positioning, guided by the Belt and Road Initiative, Guangzhou Metro positions itself as a provider of holistic solutions for global urban rail transit. Regarding market layout, it adopts a gradual strategy of "consolidating Southeast Asia, expanding into Africa, and making breakthroughs in Europe and America," having successfully delivered specific projects in countries including Singapore, Vietnam, Nigeria, and Pakistan. In terms of business content, it has notably evolved from initial single operation and maintenance services to gradually encompassing high-value-added segments such as design consulting, technology transfer, and talent development, achieving a transformation and upgrade from "selling products" to "selling services" and further to "selling standards."

Table 24-1 Overview of Guangzhou Metro Group's International Cooperation Projects

Host Country	Project Type	Specific Content / Scope	Contract Year	Project Characteristics / Significance
Ho Chi Minh City, Vietnam	Design General Contract	Feasibility Study and Preliminary Design for the Ben Thanh - Tham Luong Section of Metro Line 2	2025	Represents the first overseas general contract design project secured by GZMTR Group, marking a significant milestone in its international expansion.
Nigeria	Operation & Maintenance	Operation and Maintenance Services for Phase 1 of the Abuja Light Rail	2024	Successfully pioneered the African market, establishing a crucial foothold for future projects on the continent.
Pakistan	Operation	Operation Services for	2020	Marked the first instance of

	Service	the Lahore Orange Line Metro		GZMTR's operational business expanding internationally, demonstrating the exportability of its operational model.
Singapore	Technical Cooperation	Operations Management Consulting Project	2024	Signified a strategic partnership with a subsidiary of SMRT Corporation, facilitating knowledge exchange and collaboration with an established global operator.
Türkiye	Strategic Cooperation	Signed a Cooperation Agreement with Istanbul Metro	2024	A strategic move to explore and develop opportunities within the European market, enhancing GZMTR's global footprint.

The core competitive advantages underpinning Guangzhou Metro's internationalization lie in its comprehensive full-industry-chain service capabilities and extensive systems integration expertise. Through over three decades of development, Guangzhou Metro has established a complete industrial chain encompassing planning and design, equipment manufacturing, and operation and maintenance. In a significant move to consolidate this strength, Guangzhou Metro Group, in October 2017, jointly initiated the establishment of the Guangzhou Rail Transit Industry Alliance alongside partners such as Patech and Baiyun Electric. This alliance brings together member organizations spanning the entire value chain, including planning and design, construction, equipment manufacturing, operation and value-added services, and financial services, substantially enhancing the collective competitiveness of the industrial cluster. This integrated, cluster-based development model provides robust support for Guangzhou Metro's international endeavors, enabling it to offer customized, integrated solutions tailored to the specific needs of different countries.

Regarding innovation investment, Guangzhou Metro places high importance on the role of scientific and technological innovation in supporting its international competitiveness. By the end of 2024, Guangzhou Metro Group had invested approximately 641 million RMB in scientific and technological innovation. These investments were primarily concentrated in cutting-edge fields such as intelligent maintenance, smart stations, and energy saving and environmental protection, providing a solid technological foundation for the expansion of its international business. Furthermore, Guangzhou Metro actively promotes the international application of its innovative achievements. For instance, it has adapted and deployed intelligent maintenance technologies, proven effective in the Chinese market, for international markets based on local conditions, achieving commendable results. This approach of localizing proven innovations for global application has proven highly successful.

24.2 International Vocational Education Cooperation Models

After years of exploration and practice, Guangzhou Metro has established a multi-level, multi-dimensional international vocational education and training (VET) cooperation model system that integrates online and offline methods. This system is capable of flexibly meeting the diverse needs of different countries, different projects, and different stages.

(1) In-China Training: Immersive and Systematic Cultivation

In-China training serves as the cornerstone of Guangzhou Metro's international VET system, designed to provide partner countries with the most comprehensive and profound immersive learning experience. Its main features are as follows:

- Reliance on Top-Tier Training Bases: Training is primarily conducted at the Guangzhou Metro Training College and the "National Engineering Laboratory for System Safety and Operational Assurance of Urban Rail Transit." Trainees have access to various simulation equipment synchronized with Guangzhou Metro's actual operational environments, including full-function simulator cabs, CBTC signal simulation systems, platform screen door and control center linkage experimental platforms, and vehicle bogie maintenance training benches. This approach of "learning real technology in a real environment" significantly enhances training effectiveness.

- Systematic and Comprehensive Curriculum: The training courses are not scattered technical lectures but are systematically designed based on job competency models. The curriculum covers all levels, from senior management to frontline operations, including:

♦ Strategic Management: Network planning, Transit-Oriented Development (TOD) concepts, investment and financing models, sustainable development strategies.

♦ Operations Management: Dispatch command, passenger transport organization, emergency management, service quality control.

♦ Technical Skills: Maintenance, repair, and fault handling for majors such as rolling stock, signaling, communications, power supply, and mechanical & electrical systems.

♦ Safety Management: Safety management systems, risk assessment, accident investigation and analysis.

- Tripartite Model: "Theory + Practice + On-site Shadowing": The training adopts a progressive teaching model of "theoretical learning → simulated practice → on-site shadowing." After completing classroom learning and simulation training, trainees are assigned to Guangzhou Metro's operational lines, depots, and control centers for on-site shadowing internships. They are guided by experienced Guangzhou Metro mentors in "one-to-one" or "one-to-many" formats, personally participating in actual workflows to achieve the transformation from knowledge to capability.

- Cultural Integration and Experience: During the training period, Guangzhou Metro also organizes for trainees to visit Guangzhou's urban construction projects and experience Chinese culture, promoting cross-cultural understanding and exchange, thereby enriching technical training with humanistic warmth.

(2) Overseas Customized Training: Precision Door-to-Door Service

Aimed at trainees who cannot leave their posts for extended periods to study in China or who require on-site adaptive training for specific projects, Guangzhou Metro has launched the "Overseas Customized Training" model.

- "One-Country-One-Policy" or even "One-Line-One-Policy": Guangzhou Metro's training team conducts in-depth needs assessments in the host country before or during project implementation, thoroughly understanding the local culture, legal environment, technical foundation, and project specifics. Based on this, tailored training programs are developed. For example, for projects in countries with heavy rainfall and flooding, the training emphasizes flood emergency drills; for projects in hot and dry regions, it includes specialized courses on air conditioning systems and equipment cooling maintenance.

- Mobile Podium and On-site Teaching: Guangzhou Metro has formed an "Overseas Lecturer Corps" composed of senior engineers, technicians, and management experts. This team travels to project sites with necessary training aids, simulation software, and technical documentation. Training is conducted directly on the production front lines, such as depots, stations, and control centers, achieving "learning while observing, explaining, and practicing," thereby addressing the most practical and urgent technical and management issues.

- "Teaching How to Fish" – Trainer Cultivation: A core task in overseas customized training is cultivating local trainers ("Train the Trainer"). Guangzhou Metro experts not only transfer technology but also focus on imparting teaching methods and curriculum design concepts, helping the host country establish its own teaching faculty, thereby achieving self-sustaining and sustainable training capacity.

(3) Online Vocational Education Platform: Borderless Knowledge Sharing through Digitalization

To overcome time and space constraints and achieve low-cost, wide-coverage, and sustainable VET services, Guangzhou Metro vigorously develops its online VET platform.

- Bilingual Smart Learning Platform: Guangzhou Metro has developed or co-developed a multilingual (primarily Chinese and English) smart VET platform that integrates online courses, virtual simulations, interactive communities, and intelligent assessment. The platform content is diverse, including recorded videos, live lectures, 3D animations, and interactive fault diagnosis simulations, meeting the needs for fragmented and personalized learning.
- Application of Digital Twin and VR/AR Technologies: The platform integrates virtual stations, virtual depots, and virtual trains based on digital twin technology. Trainees can use VR equipment to conduct high-risk or high-cost operational training, such as equipment disassembly/assembly and emergency drills, in an immersive manner. For instance, trainees can simulate handling a traction system fault in a virtual environment, with the system providing real-time feedback on the correctness of their operations, significantly enhancing training safety and efficiency.
- Building Online Learning Communities: The platform builds a communication bridge between global trainees and Guangzhou Metro experts. Trainees can ask questions at any time, experts answer online, and trainees from different countries can exchange experiences, forming a continuously evolving "rail transit knowledge community."

(3) University-Enterprise Cooperative Development: Cultivating the Local Education Ecosystem

This is the most long-term and strategic cooperation model, aiming to build a rail transit talent supply chain from the source in the host country.

- Major and Curriculum Co-development: Guangzhou Metro cooperates with renowned universities or vocational colleges in the host country to jointly establish majors or colleges related to "Urban Rail Transit." Guangzhou Metro dispatches experts to participate in curriculum system design, textbook compilation, and provides teaching for some core courses.

- Training Center Co-development: The partners cooperatively build training laboratories that comply with Chinese standards and technical specifications, such as simulator driving labs, signaling control labs, and vehicle maintenance labs. Guangzhou Metro provides equipment standards, technical guidance, and sometimes even donates key equipment, ensuring that students are "job-ready upon enrollment and skilled upon graduation."
- Faculty Exchange and Joint Research: A teacher exchange mechanism is established, with Chinese teachers going abroad to teach and foreign teachers coming to Guangzhou for further study. Simultaneously, joint university-enterprise research is conducted around specific technical challenges in local rail transit development, feeding the research results back into teaching, forming a virtuous cycle of "industry-university-research."

These four models are not mutually exclusive but are often combined based on project needs, forming an international vocational education ecosystem characterized by complementary advantages and synergistic effects.

24.3 Case Study: The Vocational Education Project for the Ho Chi Minh City Metro

24.3.1 Project Background and Strategic Significance

The Ho Chi Minh City Metro Project in Vietnam represents a landmark achievement within Guangzhou Metro's internationalization strategy. In 2025, an international consortium led by Guangzhou Metro Design & Research Institute Co., Ltd. officially signed the contract for the "Project for the Adjusted Feasibility Study Report Preparation, Preliminary Engineering Design, and Procurement Services for the Ben Thanh – Tham Luong Section of the Ho Chi Minh City Urban Rail Transit." The total contract value was 1,754 billion VND (approximately 47 million RMB). This project not only signifies Guangzhou Metro's successful entry into the

Vietnamese market but also marks the first overseas rail transit overall design contract secured by Guangzhou Metro Design & Research Institute, holding ground-breaking significance in the "Going Global" process of China's urban rail transit industry.

The strategic importance of this project is reflected across multiple levels. From a market expansion perspective, Vietnam, as a rapidly developing Southeast Asian nation, exhibits strong demand for urban rail transit construction. Ho Chi Minh City is Vietnam's largest port city and economic center, and also an international friendly cooperation and exchange city of Guangzhou. Currently, Ho Chi Minh City is significantly accelerating its urban rail construction, planning to build 7 urban rail lines totaling approximately 355 kilometers by 2035. The successful implementation of this project will earn Guangzhou Metro a favorable reputation in Vietnam and the broader Southeast Asian market, laying a solid foundation for subsequent project development.

From a technology export perspective, this project signifies Guangzhou Metro's transformation and upgrade from exporting single technologies to providing integrated solutions. Within the project, Guangzhou Metro Design & Research Institute, as the overall contractor, is responsible not only for specific design work but also for coordinating and managing consortium members including Vietnam's TEDIS Company, PCC Company, and France's ARTELIA Company. The experience gained in this international collaborative design management model is of great significance for enhancing Guangzhou Metro's international capabilities.

24.3.2 Technology Transfer Content and Innovative Application

In the Ho Chi Minh City Metro Project, Guangzhou Metro is exporting not merely specific technical solutions, but an integrated design philosophy and systemic solutions infused with Chinese experience. Regarding route planning, the first phase of Ho Chi Minh City Urban Rail Transit Line 2 (Ben Thanh – Tham Luong Section) is approximately 11.3 kilometers long, comprising 9.3 kilometers of underground section and 2 kilometers of elevated section, with 11 stations in total. Leveraging its extensive experience in metro construction under complex geological conditions in Guangzhou, and tailored to the geographical and geological characteristics of Ho Chi

Minh City, Guangzhou Metro is providing comprehensive technical support encompassing geotechnical engineering, underground structures, track engineering, and electromechanical systems.

It is particularly noteworthy that Guangzhou Metro innovatively introduced the Transit-Oriented Development (TOD) model into the project. In 2025, the Department of Agriculture and Environment of Ho Chi Minh City held an exchange meeting with Guangzhou Metro Group, specifically focusing on in-depth discussions around the TOD model. The Ho Chi Minh City side explicitly expressed its desire to learn from Guangzhou Metro's mature experience in land planning, resource integration, and development utilization to help advance key rail transit projects like Metro Line 2 (Ben Thanh – Tham Luong) and related urban development processes. This indicates that Guangzhou Metro's technology export is extending beyond pure engineering construction into higher value-added areas like station-city integrated development.

In terms of the consortium cooperation model, Guangzhou Metro also demonstrated a high level of international operational capability. The consortium, with Guangzhou Metro Design & Research Institute at its core, united Vietnam's TEDIS Company, PCC Company, and France's ARTELIA Company. This cooperation model of "Chinese Technology + Localized Implementation + International Experience" ensures the project's technical standards while accommodating localization needs and an international perspective, serving as a crucial guarantee for the project's successful implementation. Through this model, Guangzhou Metro both exports Chinese technology and learns from international experience, achieving a two-way win-win situation.

24.3.3 Cooperation Model and Localization Strategy

The success of Guangzhou Metro's project in Vietnam can be largely attributed to the diversified cooperation model and deep-level localization strategy it employed. Regarding the cooperation mechanism, the project builds upon a solid foundation of prior collaboration between multiple parties. In 2023, the China Association of Metros, Guangzhou Metro Group, and the Ho Chi Minh City Management Authority for Urban Railway (MAUR) signed a Memorandum of Cooperation (MoC) in

Guangzhou, agreeing to jointly set common goals to support the research and implementation of the development strategy for Ho Chi Minh City's urban railway system. This high-level cooperation mechanism provided strong policy support and coordination guarantees for the project's specific implementation.

In terms of localization strategy, Guangzhou Metro adopted a multi-level, systematic approach. On one hand, cooperation with local enterprises through the consortium format ensured the project's full understanding and adaptation to the local market environment, laws and regulations, and technical standards. On the other hand, it actively promoted the localization of technical standards and talent cultivation. Through knowledge sharing and cooperation in personnel training, it helped enhance the local technical capacity of Ho Chi Minh City. This "teaching how to fish" strategy not only benefits the project's sustainable development but has also won the trust and recognition of the local government and enterprises.

Furthermore, targeting the characteristics of the Vietnamese market, Guangzhou Metro adopted a long-term cultivation and gradual deepening market strategy. As early as 2024, Guangzhou Metro conducted large-scale international exchange events like the "2024 Ho Chi Minh City Urban Rail Transit (Guangzhou) Training Program," successfully converting Vietnamese market opportunities. Targeting the Vietnamese market, multiple reciprocal visits by various levels of Guangzhou entities gained full recognition from different levels of the Vietnamese government. This led to the signing of intentions for cooperation in personnel training and TOD-related fields, continuously exporting Guangzhou Metro's construction, operation, and management philosophies. The "Guangzhou Model" has become a benchmark for Ho Chi Minh City's future metro development. This model, starting with talent cultivation and gradually expanding to technical cooperation and project collaboration, has laid a solid foundation for Guangzhou Metro's long-term development in the Vietnamese market.

24.4 Outcomes and Impact of Vocational Education

After more than a decade of meticulous effort, Guangzhou Metro's international vocational education initiatives have yielded substantial results. Their impact is demonstrated by the following data and facts.

24.4.1 Scale and Reach of Talent Cultivation

To date, Guangzhou Metro has trained over 10,000 rail transit professionals for more than 20 countries worldwide. The trainee base covers the complete talent spectrum, from strategic decision-makers to frontline operators.

24.4.2 Promoting the 'Going Global' of Chinese Standards

Through its vocational education exports, Guangzhou Metro has successfully promoted a wide range of Chinese rail transit technical standards, operational maintenance procedures, and management concepts overseas. The Lahore Orange Line in Pakistan fully adopts Chinese standards, while Chinese design specifications and operational standards are widely implemented in the Ho Chi Minh City Metro project.

24.4.3 Enhancing Project Success Rates and Brand Reputation

Systematic vocational education support has significantly reduced the operational risks of overseas projects. For instance, the Lahore Orange Line in Pakistan, operated and maintained by Guangzhou Metro, achieves punctuality and fulfillment rates both exceeding 99%, making it a flagship project of the China-Pakistan Economic Corridor.

24.4.4 Building a Sustainable International Cooperation Ecosystem

Vocational education cooperation has strengthened trust and connections between Guangzhou Metro and host country governments, enterprises, and academic institutions. This 'soft connectivity,' based on talent development, creates limitless possibilities for future collaboration. Guangzhou Metro has become a trusted 'lifelong partner' and capacity builder.'

Table 24-2: Guangzhou Metro International Vocational Education Output: Key Achievement Indicators

Outcome Dimension	Key Indicator	Achievements and Impact
Talent Cultivation	Number of Countries Covered	>20 countries

	Cumulative Number of Trainees	>10,000 person-times
	Localized Teacher Training	Hundreds of "seed" lecturers
Standard Output	Overseas Projects Adopting Chinese Standards	Lahore Orange Line in Pakistan, Ho Chi Minh City Metro in Vietnam, etc.
	Leading/Participating in Formulating Vocational Education Standards	Multiple industry and enterprise standards
Project Effectiveness	Key Operational Indicators of Trained Projects (e.g., Punctuality Rate)	>99% (Lahore Orange Line)
	Project Safety Incident Rate	Significantly reduced
Brand & Ecosystem	International Partner Institutions/Organizations	Covering Southeast Asia, South Asia, Africa, and other regions
	Awards/Recognition from Host Governments/Institutions	Multiple honors and recognitions

Looking ahead, through continuously innovating collaboration models, deepening the application of digital technologies, and building a global vocational education alliance, Guangzhou Metro is poised to play an even more critical role in cultivating international rail transit talent and advancing the industry as a whole.

Chapter 25 Prospects for China's Rail Transit Vocational Education Export Pathways

25.1 Strategic Context and Significance of International Vocational Education Export

With the rapid development of China's rail transit technology and the acceleration of its internationalization process, the export of China's rail transit vocational education has become a vital component of national strategy. Globally, urban rail transit is experiencing an unprecedented wave of development, particularly creating an urgent demand for talent in rail transit construction, operation, and maintenance among countries along the Belt and Road Initiative. After decades of technological accumulation and innovation, China's rail transit industry has established globally leading technical standard systems and talent cultivation models, laying a solid foundation for the international export of its vocational education.

As a leading enterprise in China's urban rail transit industry, Guangzhou Metro has accumulated extensive experience through its internationalization process. Its global footprint now spans multiple countries and regions, evidenced by projects such as the Overall Design and General Contract project for Ho Chi Minh City Metro Line 2 in Vietnam, the operation services for the Lahore Orange Line Metro in Pakistan, and the supervision of the Electric Multiple Units for the Lagos Blue Line Light Rail in Nigeria. The successful implementation of these projects not only demonstrates the advancement of Chinese rail transit technology but also highlights the critical role of vocational education and talent development in international cooperation.

The strategic significance of China's rail transit vocational education export is primarily manifested across three dimensions: the export of technical standards, the export of talent cultivation systems, and the export of cultural soft power. Through vocational education exports, China can systematically transfer its rail transit technical standards, management experience, and operation/maintenance models to partner countries, assisting them in establishing localized talent cultivation systems. Simultaneously, this enhances China's discourse power and influence within the international rail transit sector. Furthermore, exporting vocational education can drive the international development of the entire rail transit industry chain, creating a virtuous cycle integrating "technology + talent + industry."

Table 25-1 Analysis of the Strategic Value of China's Rail Transit Vocational Education Export

Strategic Dimension	**Core Content**	**Expected Benefits**
Technical Standards Export	Complete standard system for rail transit construction, operation, and management	Enhance international discourse power, promote the internationalization of Chinese standards
Talent Cultivation Export	Curriculum systems, practical training models, certification standards	Cultivate pro-China talent, promote localized capacity building
Industrial Chain Export	Entire industrial chain covering	Drive coordinated global expansion of the

	planning/design, equipment manufacturing, operation/maintenance	industrial chain, create comprehensive value
Cultural Soft Power Export	Craftsman spirit, professional competence, management philosophy	Strengthen cultural influence, enhance national image

25.2 Practices of Vocational Education Export in Guangzhou Metro's Internationalization Process

25.2.1 International Application of the Customized Talent Training Model

In its international development, Guangzhou Metro innovatively applied the domestically mature customized training class model to international talent cultivation. In 2019, Guangzhou Transportation Technician Institute signed a joint training agreement with Guangzhou Metro Group Co., Ltd., successfully establishing the 2021 session of the Guangzhou Metro customized training class. This demand-oriented model involves collaborative development of training plans, curriculum systems, and dual-qualified teaching teams between the enterprise and the institute, achieving seamless integration between talent training and job requirements.

In international applications, Guangzhou Metro has adaptively modified this model. Tailoring to the cultural backgrounds, education systems, and talent demand characteristics of different countries, it has developed localized training plans for customized classes. Taking the Ho Chi Minh City project in Vietnam as an example, Guangzhou Metro collaborated with local educational institutions to jointly cultivate technical and skilled personnel meeting the development needs of Vietnam's rail transit. Through the approach of "joint enrollment, joint training, and joint assessment" by the enterprise and institute, the relevance and practicality of talent training were ensured, providing talent support for the long-term stable operation of the cooperative project.

25.2.2 International Cooperation Model Integrating Industry, Education, and Research

In its international vocational education export, Guangzhou Metro emphasizes deep integration of industry, education, and research, forming diversified international cooperation models. On one hand, Guangzhou Metro actively promotes cooperation with overseas universities and research institutions. For instance, it established a joint postgraduate training mechanism with the Hong Kong University of Science and Technology, jointly cultivating PhD students in three directions: demand prediction and planning for urban agglomeration transportation systems based on big data, AI-enabled rail transit operation management, and integrated transportation information sharing with multi-modal coupling. This high-level talent training model provides strong talent and intellectual support for the international development of the rail transit industry.

On the other hand, Guangzhou Metro integrates industrial chain resources through industrial alliances, enhancing the systematicness and completeness of vocational education export. In October 2017, Guangzhou Metro Group, jointly with Potevio Group and Baiyun Electric, initiated the establishment of the Guangzhou Rail Transit Industry Alliance. Member units cover the entire industrial chain, including planning and design, construction, equipment manufacturing, operation and value-added services, and financial services. This full industrial chain collaboration model provides abundant resource support and practical platforms for international vocational education export, enabling trainees to comprehensively understand the technical requirements and management specifications of all aspects of rail transit.

25.2.3 Development of International Curriculum Systems and Training Standards

Based on extensive international project experience, Guangzhou Metro has developed a set of curriculum systems and training standards adapted to international needs. This system, based on Chinese rail transit technical standards, fully considers factors such as the cultural characteristics, climate conditions, and operation and maintenance habits of different countries, forming a modular and customizable course structure. The course content covers the entire lifecycle of rail transit, including planning and design, engineering construction, operation management, and equipment maintenance, meeting the training needs of personnel at different levels and in different positions.

Regarding training standards, Guangzhou Metro draws on advanced domestic vocational education concepts, such as the integrated education model of "post requirements, curriculum design, skill competitions, and certification," combining them with international requirements to develop a comprehensive training system integrating skill training, qualification certification, and career development. Simultaneously, targeting the characteristics of countries along the Belt and Road Initiative, Guangzhou Metro has also developed specialized "Chinese language + vocational skills" training programs. These not only impart professional skills but also promote language and cultural exchange, enhancing the training effectiveness.

25.3 Building a Multi-stakeholder Collaborative Model for Vocational Education Export

25.3.1 Government-Enterprise-Institution Collaborative Export Model

China's rail transit vocational education export requires the establishment of a multi-stakeholder collaborative mechanism guided by the government, led by enterprises, and with the participation of educational institutions.

At the government level, policy support, diplomatic channels, and financial assistance create a favorable external environment for vocational education export. For instance, the National Rail Transit Equipment Industry-Education Integration Consortium, supported by the Ministry of Education, is a key government-guided initiative. It aims to integrate high-quality industry resources, align industry demands with educational supply, and provide talent support for the "high-end, intelligent, green, and international" development of the rail transit equipment industry.

At the enterprise level, leading companies like Guangzhou Metro, leveraging their extensive engineering experience and mature technical management systems, provide practical and systematic training content and practical training platforms for vocational education export. Enterprises drive this export through overseas projects, conducting localized talent cultivation during project implementation. This approach not only addresses the project's operational manpower needs but also cultivates the technical and skilled personnel required for the long-term development of partner countries.

At the institution level, vocational colleges and universities fully utilize their educational resources and teaching expertise to cooperate with enterprises in developing internationalized curricula and teaching materials, co-building training bases, and jointly conducting talent cultivation. For example, Guangzhou Railway Polytechnic collaborates with leading enterprises like Guangzhou Railway Group and Guangzhou Metro to advance the integrated education model of "Post-Class-Competition-Certification." It also jointly initiated the establishment of the National Rail Transit Safety Control Industry-Education Integration Consortium, focusing on key areas such as green rail transit and rail transit signal control safety.

25.3.2 Remote Training Model Based on Digital Technology

With the development of information technology, digital and intelligent remote training models are becoming an important direction for rail transit vocational education export. The application of technologies such as virtual simulation, augmented reality, and cloud computing can break time and space constraints, enabling the global sharing of educational resources and intelligent management of the training process.

Specifically, a Rail Transit Cloud Training Platform can be constructed. This platform digitally models complex training content such as equipment structure, operational procedures, and fault handling, allowing for visual demonstration and interactive operation through virtual simulation. This enables trainees to master core skills without being limited by physical equipment. Simultaneously, using big data and artificial intelligence technologies to analyze trainees' learning behaviors can provide them with personalized learning paths and tutoring plans, thereby improving training efficiency.

Furthermore, utilizing Digital Twin technology, a virtual rail transit system fully corresponding to the physical system can be built. This enables training functions such as real-time system status monitoring, fault prediction and diagnosis, and emergency response drills, enhancing trainees' systems thinking and problem-solving abilities. This digital training model can not only reduce training costs and improve efficiency but also decrease reliance on physical equipment, making it more suitable for promotion in developing countries with relatively weak infrastructure.

25.3.3 Tiered and Classified Precision Export Strategy

A tiered and classified precision strategy should be adopted for China's rail transit vocational education export, tailored to the characteristics of different countries and regions. Based on factors such as the economic development level, rail transit development stage, and educational foundation of partner countries, the target markets can be categorized into three types: advanced markets, growth markets, and emerging markets.

For advanced markets, represented by developed countries like Singapore, the focus of vocational education export should be on high-end technology and management training. This includes cutting-edge areas such as intelligent maintenance, green and energy-saving technologies, and networked operations. The aim is to achieve mutual complementarity of advantages and jointly promote industry development through technological exchange and collaborative research.

For growth markets, represented by middle-income countries like Vietnam and Pakistan, vocational education export should emphasize systematic technology application and operation and maintenance management training. The goal is to assist these countries in establishing a comprehensive rail transit talent cultivation system, including major setup, curriculum development, and training base construction, to support the sustainable development of their rail transit networks.

For emerging markets, represented by low-income countries like Nigeria, vocational education export should primarily focus on basic skills and operational maintenance training. The emphasis should be on the practicality and adaptability of the courses, helping to cultivate the fundamental technical and skilled personnel required for rail transit operations, thereby laying the foundation for the development of their rail transit industry.

Table 25-2 Comparison of Vocational Education Export Strategies for Different Market Types

Market Type	Representative	Export Focus	Cooperation Mode	Expected Goals

	Countries/Regions			
Advanced Markets	Singapore, European countries	High-end technologies, Management innovation, Joint standard development	Technical exchanges, Joint R&D, Faculty exchanges	Enhance international influence, Achieve technological mutual benefit
Growth Markets	Vietnam, Pakistan, Kazakhstan	Systematic technology application, O&M management, Talent cultivation systems	Industry-academy cooperation, Customized training classes, Localized adaptation	Expand market share, Establish long-term cooperation
Emerging Markets	Nigeria, Other African countries	Basic skills, Operation & maintenance, Hands-on training	Government-aided projects, Project-driven initiatives, Training base construction	Cultivate future markets, Build brand image

25.4 Implementation Pathways and Strategic Recommendations for Vocational Education Export

25.4.1 Curriculum Development Integrating Standardization and Localization

The success of China's rail transit vocational education export hinges on achieving an organic unity of standardization and localization. On one hand, it is essential to adhere to Chinese technical standards as the foundation, developing systematic and standardized curriculum systems to ensure the professionalism and reliability of the training content. On the other hand, full consideration must be given to the cultural background, legal regulations, and climate conditions of the target country, making adaptive adjustments to the course content to enhance its relevance and practicality.

In curriculum development, one can draw on the experience of the apparel major reform at Wuxi Vocational Institute of Arts & Technology. In response to the development of new knowledge, technologies, and processes in the industry, it innovatively reconstructed "modular skill" training components, clearly dividing vocational skills into two layers: basic standards and post-specific standards. This involved building standardized course modules and dynamically updated modules to ensure the teaching content keeps pace with cutting-edge industrial technologies. Simultaneously, by conducting in-depth visits to leading enterprises and analyzing the competency requirements of core industry positions, tailored and precise curriculum solutions can be developed.

Furthermore, curriculum development should emphasize modular design, forming course packages for different specializations and skill levels, facilitating flexible combination according to the needs of different countries. The course content should reflect the integration of theory and practice, be equipped with rich case studies and practical assignments, and be presented in digital formats for easy dissemination and updating.

25.4.2 Enhancing the International Competence of Dual-Qualified Teaching Staff

A high-quality, dual-qualified teaching staff is a crucial guarantee for the quality of vocational education export. Dual-qualified teachers must possess not only solid theoretical foundations and teaching abilities but also rich engineering practical experience and cross-cultural communication skills. To this end, a systematic teacher training mechanism needs to be established to continuously improve teachers' international competence and professional level.

On one hand, a mechanism for teachers to undertake enterprise practice should be established, regularly selecting and sending teachers to the front lines of domestic and international rail transit projects for practical training. This allows them to understand the latest technological developments and industry demands, accumulating engineering practical experience. For example, Guangzhou Metro Group supports CRRC and member enterprises of the consortium in establishing national vocational education teacher enterprise practice bases and "Dual-Qualified" teacher training bases. These focus on developing teacher learning and practice programs around new knowledge, technologies, processes, materials, equipment, and standards in the industry.

On the other hand, the cultivation of teachers' cross-cultural teaching abilities must be strengthened. Through international exchanges, language training, and cultural immersion, teachers' foreign language proficiency and cross-cultural communication skills can be improved, enabling them to adapt to the international teaching environment. Concurrently, senior enterprise experts and foreign teachers can be invited to join the teaching team, forming an internationalized teaching staff that combines full-time and part-time members and integrates Chinese and foreign expertise.

25.4.3 Co-construction, Sharing, and Digital Upgrading of Training Bases

Training bases are essential carriers for rail transit vocational education export and play an irreplaceable role in cultivating skilled talents. In international vocational education export, efforts should be actively made to promote the co-construction, sharing, and digital upgrading of training bases to improve resource utilization efficiency and training effectiveness.

On one hand, drawing on the experience of the national industry-education integration consortium in constructing national-level industry-education integrated training bases, member units of the consortium can be supported in building a number of high-level training bases categorized by specialty and field. These should cover training functions for rail transit equipment such as high-speed EMUs, urban rail vehicles, and high-power locomotives, constructing typical scenarios including trial production and testing, manufacturing, intelligent maintenance, upkeep services, and emergency drills related to rail transit equipment. These training bases can be built either domestically, attracting overseas trainees to China for training, or overseas, directly serving local talent cultivation.

On the other hand, the development and application of virtual simulation training resources should be strengthened. Through technologies like digital twins and virtual reality, virtual training platforms corresponding to physical training bases can be constructed, achieving visualization and repeatability of the training process, thereby reducing training costs and risks. For instance, Nanning No.1 Vocational Technical

School built an autonomous region-level demonstrative virtual simulation training base for vocational education, providing students with a safer, more flexible, and efficient training environment.

25.4.4 Construction of Qualification Certification and Quality Assurance Systems

Establishing an internationally recognized vocational qualification certification system is a crucial step towards standardization and normalization for China's rail transit vocational education export. By cooperating with internationally authoritative certification bodies and developing a vocational qualification certification system based on Chinese technical standards and job competencies, authoritative talent evaluation standards can be provided to partner countries, enhancing the credibility and influence of vocational education export.

Specifically, focusing on key rail transit positions, their job responsibilities and competency requirements can be analyzed to formulate corresponding skill level standards and assessment methods, forming a systematic vocational qualification certification framework. Simultaneously, establishing certification processes and management systems aligned with international standards ensures the standardization of the certification process and the comparability of certification results.

Furthermore, a comprehensive quality assurance system needs to be constructed to monitor the quality of the entire vocational education export process and enable continuous improvement. This includes formulating teaching quality standards, establishing multi-stakeholder evaluation mechanisms, and forming closed-loop feedback for improvement, ensuring that vocational education export activities consistently maintain high quality. For example, Nanning No.1 Vocational Technical School effectively addressed the disconnection between talent cultivation and industry demands through its "Five Integrations, Three Connections, and Two Extensions" industry-education integrated new-quality skilled talent cultivation system, providing a strong guarantee for talent cultivation quality.

25.5 Safeguard Mechanisms and Future Prospects for Vocational Education Export

25.5.1 Policy Support and Resource Integration Mechanisms

The sustainable development of China's rail transit vocational education export requires strong policy support and efficient resource integration mechanisms.

At the government level, policy support should be further intensified by incorporating vocational education export into the national framework for foreign cooperation. Promoting the signing of educational cooperation agreements with key countries through diplomatic channels will create a favorable policy environment. Simultaneously, establishing special funds for vocational education export to provide financial support to eligible projects and institutions can reduce export costs and enhance motivation.

At the industry level, industry organizations should fully play their role as bridges and bonds, strengthening industry self-discipline and standard construction, and promoting information sharing and resource integration. For example, the China Association of Metros played a significant role in the cooperation between Guangzhou Metro Group and the Ho Chi Minh City Management Authority of Urban Railway (MAUR). The three parties signed a Memorandum of Cooperation, working together to set common goals to support research and implementation of the development strategy for Ho Chi Minh City's urban railway system.

At the enterprise level, leading enterprises like Guangzhou Metro should further improve their internal mechanisms, integrate vocational education export into their internationalization strategy, and establish dedicated departments and teams responsible for the planning, implementation, and management of vocational education export. Concurrently, enhancing collaboration with upstream and downstream enterprises in the industrial chain will form a synergistic force for vocational education export.

25.5.2 Risk Prevention and Sustainable Development Mechanisms

Vocational education export in rail transit faces numerous risks, including political risks, cultural conflicts, and quality risks, necessitating the establishment of systematic risk prevention and control mechanisms. In the project's preliminary phase, thorough market research and risk assessment should be conducted to understand the target country's political situation, laws and regulations, and cultural practices, providing a basis for project decision-making. During project implementation, risk monitoring and early warning mechanisms should be established to promptly identify and address potential risks, ensuring the project progresses smoothly. Simultaneously, respecting the cultural traditions and development needs of the target country and seeking mutually beneficial cooperation models are essential for achieving the sustainable development of vocational education export.

Specifically, lessons can be drawn from the experience of Tianjin Rail Transit Group signing a cooperation agreement with the Astana Municipal Government in Kazakhstan. Upholding the Silk Road spirit of "peace and cooperation, openness and inclusiveness, mutual learning, and mutual benefit," they fully invested their rail transit construction and operation experience into the project, helping to achieve a leapfrog improvement in the level of local urban transport infrastructure. This concept of mutually beneficial cooperation is the fundamental guarantee for the sustainable development of vocational education export.

25.5.3 Future Development Trends and Prospects

With the rapid development of global rail transit and the deepening of China's internationalization process, China's rail transit vocational education export will embrace broader development space. In the future, China's rail transit vocational education export will exhibit the following trends:

Firstly, the export content will become more comprehensive, shifting from purely technical and skill training to integrated exports encompassing technical standards, management concepts, and professional culture, forming an all-encompassing vocational education export system.

Secondly, the export methods will become more digitalized. The application of technologies such as virtual simulation, remote interaction, and intelligent assessment will become more widespread, and blended training models integrating online and offline elements will become mainstream.

Thirdly, cooperation models will become more diversified, evolving from one-way export to two-way exchange, and deepening from project cooperation to system building, forming a new pattern of international cooperation that is more equal, open, and inclusive.

Fourthly, quality assurance will become more standardized. Internationally recognized vocational qualification certification systems and quality assessment standards will be gradually established, continuously improving the quality and level of vocational education export.

Looking ahead, China's rail transit vocational education export should be guided by the vision of building a community with a shared future for mankind, adhering to the principles of extensive consultation, joint contribution, and shared benefits. Through high-quality vocational education cooperation, it will promote the cultivation of global rail transit talent and technological progress, making greater contributions to the Belt and Road Initiative and global sustainable development.

Made in the USA
Las Vegas, NV
04 November 2026

57423832R10200